SCHÄFFER
POESCHEL

KPMG AG Wirtschaftsprüfungsgesellschaft (Ed.)

ESRS – A Visual Approach

Structured Overviews of the European Sustainability Reporting Standards

1. Auflage

Schäffer-Poeschel Verlag Stuttgart

Bibliografische Information der Deutschen Nationalbibliothek

Die Deutsche Nationalbibliothek verzeichnet diese Publikation in der Deutschen Nationalbibliografie; detaillierte bibliografische Daten sind im Internet über http://dnb.dnb.de/ abrufbar.

Print: ISBN 978-3-7910-6392-8 Bestell-Nr. 11142-0001
ePDF: ISBN 978-3-7910-6393-5 Bestell-Nr. 11142-0150

KPMG AG Wirtschaftsprüfungsgesellschaft (Ed.)
ESRS – A Visual Approach
1. Auflage, Mai 2024

www.schaeffer-poeschel.de
service@schaeffer-poeschel.de

Bildnachweis (Cover): Midjourney (KI-generiert)

Produktmanagement: Anna Pietras

Schäffer-Poeschel Verlag Stuttgart
Ein Unternehmen der Haufe Group SE

Foreword

In today's complex business landscape, undertakings face increasing pressure to effectively capture, evaluate and disclose their environmental, social and governance-related data. Transparent and accurate reporting is not only crucial for sustainable business practices, but it also fosters trust among stakeholders and investors.

As a European Union directive, the CSRD has set a milestone requiring undertakings to report not only on their financial performance, but also on their environmental and social impacts. In particular, the transition from NFRD to CSRD extends the obligation to disclose non-financial information to more undertakings, while increasing the required level of detail of the information to be reported.

Therefore, it is with great pleasure that we present to you this first publication on visualising the European Sustainability Reporting Standards – ESRS.

With the collaboration of experts from diverse industries, we have endeavoured to support undertakings in fulfilling reporting requirements while enhancing the clarity and accessibility of their reporting. Our visualisation provides a clear understanding of the legal requirements and empowers an efficient implementation strategy.

With this publication, you have a tool at hand that guides you through each of the ESRS Standards in a goal-oriented way. Visual contexts make a significant contribution to the clear understanding of the detailed Disclosure Requirements and in the subsequent application. In addition to the qualitative and quantitative information of the standards we present related frameworks and tools. Definitions and references to the ESRS support profound research.

The publication is published as a first edition and addresses a broad audience from business, politics and society. We firmly believe that visualisation goes beyond enhancing the quality of ESRS information; it empowers decision-makers at all levels of an undertaking to uncover trends, gain a deeper understanding of interconnections and develop forward-thinking strategies. It provides you with ideas in the context of sustainability reporting skills and sheds light on an already very relevant as well as dynamic future topic.

Our heartfelt gratitude goes out to all employees who contributed their valuable knowledge and expertise to the creation of this publication. Their passion and dedication have enabled us to produce a comprehensive and practical resource that aims to inspire and guide you in your own ESRS reporting endeavours. A special thanks for supporting the creation of this publication goes to Johann Schnabel, Dr. WP/StB Jan-Hendrik Gnändiger, WPin Stefanie Jordan and WPin/StBin Kira Terbeck.

The Board
KPMG in Germany

Berlin, January 2024

Publication dates of the standards

Standards	Publication date
Cross-cutting standards	
ESRS 1 General requirements	December 2023
ESRS 2 General disclosures	December 2023
Environment	
E1 Climate change	December 2023
E2 Pollution	December 2023
E3 Water and marine resources	December 2023
E4 Biodiversity and ecosystems	December 2023
E5 Resource use and circular economy	December 2023
Social	
S1 Own workforce	December 2023
S2 Workers in the value chain	December 2023
S3 Affected communities	December 2023
S4 Consumers and end-users	December 2023
Governance	
G1 Business conduct	December 2023

Table of contents

List of abbreviations

AMSB	Administrative, management and supervisory bodies
Annex II	Annex II of the Delegated Regulation on the first set of ESRS (Acronyms and glossary of terms)
AR	Application Requirement
CapEx	Capital expenditure
CO2eq	Carbon dioxide equivalent
CSRD	Corporate Sustainability Reporting Directive
DR	Disclosure Requirement
E	Environment
EEA	European Economic Area
EFRAG	European Financial Reporting Advisory Group
ESRS	European Sustainability Reporting Standards
EU	European Union
FTE	Full time equivalent
G/GOV	Governance
GHG	Greenhouse gas
HSM	Health and safety management
ILO	International Labour Organisation
IROs	Impacts, risks and opportunities
ISO	International Organization of Standardization
IUCN	International Union for Conservation of Nature
kg	Kilogramme
LEAP approach	Locate, evaluate, assess, prepare approach
m^3	Cubic meter
MDR-A/P/M/T	Minimum DRs – Actions/Policies/Metrics/Targets
M	Million
NFRD	Non-Financial Reporting Directive
NGO	Non-governmental Organisation
No.	Number
OECD	Organisation for Economic Co-operation and Development
OpEx	Operational expenditure
PIEs	Public interest entities
S	Social
SDG	Sustainable Development Goal
SFDR	Sustainable Finance Disclosure Regulation
SMEs	Small and medium enterprises
t	Tons
UN	United Nations
UNGPs	UN Guiding Principles

1 Introduction

The Non-Financial Reporting Directive (NFRD) currently regulates the disclosure of non-financial information for large public interest entities (PIEs) in the EU with over 500 employees. However, a transformative change is on the horizon with the impending Corporate Sustainability Reporting Directive (CSRD) set to replace the NFRD and significantly expand non-financial reporting regulations within the EU.

The CSRD will initially target undertakings subject to the NFRD, starting with the annual reports for the financial year 2024. However, its scope will gradually increase to encompass new entrants such as large non-listed undertakings, listed small and medium enterprises (SMEs) and non-EU undertakings at the individual level and, if applicable, at its group level, ushering in a new era of sustainability reporting.

The CSRD stands as a pivotal cornerstone of the European Green Deal and the Sustainable Finance Agenda, aiming to provide comprehensive and valuable information on environmental (E), social (S) and governance (G) issues. This directive, part of EU law, seeks to harmonise and unify sustainability statements within the EU.

To achieve this harmonisation, the European Financial Reporting Advisory Group (EFRAG) has been instrumental in developing the European Sustainability Reporting Standards (ESRS). These standards will serve as delegated acts adopted by the European Commission, creating a unified framework for sustainability statements across diverse industries and sectors. The ESRS implement the content of the CSRD and will offer clear guidelines to ensure consistency and comparability in reporting practices, empowering organisations to demonstrate their commitment to sustainable development.

By enforcing equal standards for all reporting undertakings, the CSRD enhances comparability among sustainability statements, enabling stakeholders to easily assess and compare the sustainability performance of different organisations. Moreover, the publication of the sustainability statements in the management reports will increase their availability, making them readily accessible to stakeholders.

The CSRD's reach will extend beyond European undertakings, as non-EU undertakings meeting specific criteria will also be subject to the unified framework for sustainable business practices.

Through mandatory assurance reporting with limited assurance from 2024 and reasonable assurance planned from 2028, the CSRD further promotes improved data quality, enhancing the accuracy and reliability of sustainability reports.

Within the realm of sustainability reporting, the concept of double materiality gains significance. It considers two dimensions of sustainability views: impacts and risks and opportunities. Impacts represent the effects of the undertaking on people and the environment. Risks and opportunities mean the financial effects of people and the environment on the undertaking. This approach enables organisations to address sustainability issues that may have significant financial implications, while also addressing broader societal and environmental concerns.

The sector-agnostic ESRS, published in December 2023, provide a comprehensive framework applicable to organisations across all sectors. In the coming years, sector-specific standards, third-country undertaking standards and SME standards tailored to the unique characteristics of different industries or undertakings will further refine reporting requirements.

Through the CSRD and the development of the ESRS, sustainability reporting is set to become a more standardised, transparent and effective practice, enabling organisations to uphold sustainable development while meeting the expectations of stakeholders and contributing to a greener and more sustainable future.

2 General information on CSRD and ESRS

Overview of CSRD (I/I)

Timeline and scoping

Objective

Enabling understanding of:

- The comprehensive Union framework on sustainability reporting that contains an expansion of the scope of the reporting requirements to additional categories of undertakings, the introduction of an assurance requirement and the application of mandatory sustainability reporting standards.
- The sustainability information that an undertaking shall disclose in accordance with the Accounting Directive to adequately and comparably inform investors, civil society actors and other stakeholders and to close the gap between users' information needs and the sustainability information provided by the undertakings.

Scope

Undertakings or groups subject to the existing NFRD

PIEs registered in the EU, exceeding the annual average of 500 employees and meeting at least one of the following two criteria in two consecutive financial years:

- Total assets: > €25 M[(a)]
- Net turnover: > €50 M[(a)]

Increased size limits for groups

(a): In case of an assessment of groups on an unconsolidated view EU member states may have increased the limits for the total assets and net turnover criteria by 20% in national transposition of the Accounting Directive.

All large undertakings or groups not subject to the NFRD

Large undertakings subject to the Accounting Directive registered in the EU and meeting at least two of the following three criteria in two consecutive financial years:

- Annual average employees: > 250
- Total assets: > €25 M[(a)]
- Net turnover: > €50 M[(a)]

Listed SMEs and small and non-complex credit institutions and captive insurers

Listed SMEs registered in the EU and meeting at least two of the following three criteria in two consecutive financial years, with an option to defer their sustainability reporting until 2028 and with a limited reporting obligation:

- Annual average employees: > 10
- Total assets: > €450,000
- Net turnover: > €900,000

Small and non-complex credit institutions and captive insurers registered in the EU meeting at least two of the following three criteria in two consecutive financial years:

- Annual average employees: > 250
- Total assets: > €25 M[(a)]
- Net turnover: > €50 M[(a)]

Non-EU undertakings

Undertakings not registered in the EU will also be impacted by the CSRD through requirements of their EU subsidiaries or branches if:

- Net turnover of the group in the EU: > €150 M for each of the last two consecutive financial years; and large subsidiaries registered in the EU exist, or
- Net turnover of the group in the EU: > €150 M for each of the last two consecutive financial years; no large subsidiaries registered in the EU exist and net turnover: > €40 M generated by at least one EU branch during the preceding year.

Timeline and assurance

Publication of the annual reports for the financial year 2024	Publication of the annual reports for the financial year 2025	Publication of the annual reports for the financial year 2026	Publication of the annual reports for the financial year 2028
Annual reports shall be accompanied by an assurance opinion based on a limited assurance engagement			Reasonable assurance opinion planned

Overview of ESRS (I/XIII)

Structure of ESRS

Topics	Standards	Relevant additional documents
Cross-cutting standards **Cross-cutting**	**ESRS 1 –** General requirements **ESRS 2 –** General disclosures	**Annex to the Delegated Regulation on the first set of ESRS** Annex I – ESRS Annex II – Acronyms and glossary of terms
Topical standards **Environment**	**ESRS E1 –** Climate change **ESRS E2 –** Pollution **ESRS E3 –** Water and marine resources **ESRS E4 –** Biodiversity and ecosystems **ESRS E5 –** Resource use and circular economy	**EFRAG Basis for conclusions related to the draft ESRS issued in November 2022** The basis for conclusions supplement each standard by providing further details on: • Objective and context • Process and reasoning for mandating a Disclosure Requirement (DR) • Other DRs considered
Social	**ESRS S1 –** Own workforce **ESRS S2 –** Workers in the value chain **ESRS S3 –** Affected communities **ESRS S4 –** Consumers and end-users	**Relevant EFRAG Appendices to the draft ESRS issued in November 2022** Appendix II – Implementation of the CSRD in the ESRS
Governance	**ESRS G1 –** Business conduct	**Implementation guidance published by EFRAG in 2024** EFRAG IG 1 – Materiality assessment EFRAG IG 2 – Value chain EFRAG IG 3 – List of ESRS datapoints

Sector-agnostic

Sector-specific standards: Planned but not yet published

Third-country standards: Planned but not yet published

SME standards: Planned but not yet published

Entity-specific disclosures: Entity-specific disclosures apply for material IROs not or insufficiently addressed by an ESRS. The requirement for these disclosures is included in ESRS 1 General requirements.

Impacts, risks and opportunities – a driving principle in ESRS

Scope of information to be included in the sustainability statements based on material impacts, risks and opportunities (IROs)

Upstream value chain

Based on material IROs connected with the undertaking

ESRS do not require information on each and every actor in the upstream value chain but only the inclusion of material upstream value chain information.

[ESRS 1.64]

+

Own operations

Based on the undertaking

Own operations include the business activities of the undertaking or the group and its subsidiaries.

+

Downstream value chain

Based on material IROs connected with the undertaking

ESRS do not require information on each and every actor in the downstream value chain but only the inclusion of material downstream value chain information.

[ESRS 1.64]

Overview of ESRS (III/XIII)

Impacts, risks and opportunities – a driving principle in ESRS

Impacts, risks and opportunities and their relevance for sustainability reporting according to ESRS

The ESRS specify the information that an undertaking shall disclose about its **material IROs** in relation to environmental, social and governance sustainability matters. [2] IROs are defined as follows:

Impacts

Actual or potential positive or negative short-, medium- or long-term effects of the undertaking on the environment and people connected with the undertaking's own operations and upstream and downstream value chain, through its products and services, as well as through its business relationships. The impacts can be intended or unintended and reversible or irreversible.
[Annex II of the Delegated Regulation on the first set of ESRS (Annex II)]

Risks

Sustainability-related risks with negative financial effects arising from environmental, social or governance matters that may negatively affect the undertaking's financial position, financial performance, cash flows, access to finance or cost of capital over the short-, medium- or long-term.
[Annex II]

Opportunities

Sustainability-related opportunities with positive financial effects arising from environmental, social or governance matters that may positively affect the undertaking's financial position, financial performance, cash flows, access to finance or cost of capital over the short-, medium- or long-term.
[Annex II]

A complete depiction of an **IRO** includes all material information necessary for the users to understand that impact, risk or opportunity. Therefore, the undertakings shall report against the following four **reporting areas**: [QC 6]

Governance

The governance processes, controls and procedures used to monitor, manage and oversee IROs.
[ESRS 1.12 (a)]

Strategy

How the undertaking's strategy and business model (SBM) interact with its material IROs, including how the undertaking addresses those IROs.
[ESRS 1.12 (b)]

Impact, risk and opportunity management

The process(es) by which the undertaking identifies IROs, assesses their materiality and manages material sustainability matters through policies and actions. [ESRS 1.12 (c)]
IRO management includes policies (a set of frameworks or management principles to implement the strategy or management decisions) and actions (actions and action plans to ensure the delivery against targets set and to address material IROs).
[Annex II]

Metrics and targets

Metrics are qualitative and quantitative indicators that are used to measure and report on the effectiveness of the delivery of sustainability-related policies and against targets over time. Targets refer to measurable, outcome-oriented and time-bound goals that the undertaking aims to achieve in relation to its material IROs.
[Annex II]

Overview of ESRS (IV/XIII)

Impacts, risks and opportunities – a driving principle in ESRS

In order to identify the reporting contents of its sustainability statement, the undertaking needs to assess the materiality of IROs linked to so-called sustainability matters

ESRS 2

↓

Disclosure of the datapoint in any case. No assessment of IROs required.

Topical ESRS/sector-specific ESRS/entity-specific sustainability matters

↓

Assessment of the materiality of IROs linked to sustainability matters

Undertaking → Impacts → Sustainability matter

Sustainability matter → Risks and opportunities → Undertaking

VIII/XIII

ESRS 1 V/XXIX

Note:

The term "sustainability matters" refers to all topics, sub-topics and sub-sub-topics that are listed in ESRS 1 Appendix A [Application Requirement (AR) 16]. Generally, a topic covers several sub-topics as well as a sub-topic covers several sub-sub-topics. IROs can occur on all levels, e.g. on topic, sub-topic or sub-sub-topic level. For some ESRS there are no sub-sub-topics defined. The sustainability matters listed shall be supplemented by entity-specific sustainability matters.

Sustainability matters	Topic								
	Sub-topic 1			Sub-topic 2			…		
	Sub-sub-topic 1	Sub-sub-topic 2	…	Sub-sub-topic 1	Sub-sub-topic 2	…	Sub-sub-topic 1	Sub-sub-topic 2	…

Overview of ESRS (V/XIII)

Impacts, risks and opportunities – a driving principle in ESRS

The assessment of IROs linked to sustainability matters

Due diligence process and materiality assessment (ESRS 1 XIV/XXIX; ESRS 1 IX/XXIX)

Stakeholder engagement (ESRS 1 VII/XXIX)

Engagement (e.g. dialogue) with affected stakeholders is central to the undertaking's process to identify and assess actual and potential negative impacts. The undertaking may engage with affected stakeholders or their representatives (such as employees or trade unions), along with users of sustainability reporting and other experts, to provide inputs or feedback on its conclusions regarding its material IROs. However, ESRS 1 does not mandate active engagement in dialogue with affected stakeholders.

Identification of impacts (ESRS 1 IV/XXIX)

The identification of actual and potential positive or negative short-, medium- and long-term impacts from own operations and along the upstream and downstream value chain emerges from the engagement with relevant stakeholders. Impacts often trigger future risks and opportunities.

Identification of risks and opportunities (ESRS 1 IV/XXIX)

When identifying risks and opportunities the undertaking shall consider the existence of dependencies on natural and social resources as sources of financial effects and the classification as source of risks/opportunities (contributing to negative/positive deviation in future expected cash inflows or increase/decrease in deviation in future cash outflows and/or negative/positive deviation from expected change in capitals not recognised in financial statements).

Assessment of the materiality of impacts (impact materiality) (ESRS 1 X/XXIX)

Impact materiality
=
Likelihood of the impact
&
Severity of the impact

- Negative impacts: Scale & Scope & Irremediable character
- Positive impacts: Scale & Scope

Potential impacts: Likelihood of the impact & Severity of the impact; Actual impacts: Severity of the impact

+

Assessment of the materiality of risks and opportunities (financial materiality) (ESRS 1 X/XXIX)

Financial materiality
=
Likelihood of occurrence of the financial effect
&
Potential magnitude of the financial effects

Financial effects:

Risks or opportunities that have (or could reasonably be expected to have) a **material influence** on the undertaking's:

Develop-ment **or** Financial position **or** Cash flows **or** Access to finance **or** Cost of capital

Over the **short-, medium- or long-term**

Overview of ESRS (VI/XIII)

Typical structure of a topical standard

Objective	Includes the objective of the standard, related EU legislation and regulation, covered sustainability matters and their scope as well as topic-specific information.
Interaction with other ESRS	The interacting standards (ESRS 1 General requirements and ESRS 2 General disclosures) should be read in conjunction with the topical standard. Intersectional topics covered by other standards are negatively delimited.
DRs	Contains the topical DRs to be presented in the sustainability statement.
↳ **ESRS 2 General disclosures**	ESRS 2 requirements consisting of governance, strategy and IRO management are extended by topic-specific requirements.
↳ **IRO management**	Provides information and guidelines on how to identify, assess and manage the material IROs associated with that specific standard. The chapter contains policies and actions.
↳ **Metrics and targets**	Defines and establishes measurable metrics and targets related to the specific standard.
Appendix: ARs	Topic-specific information on the practical application of the DRs. ARs can include additional DRs.

[- - -] Explanation/Examples

Overview of ESRS (VII/XIII)

Typical structure of a topical standard

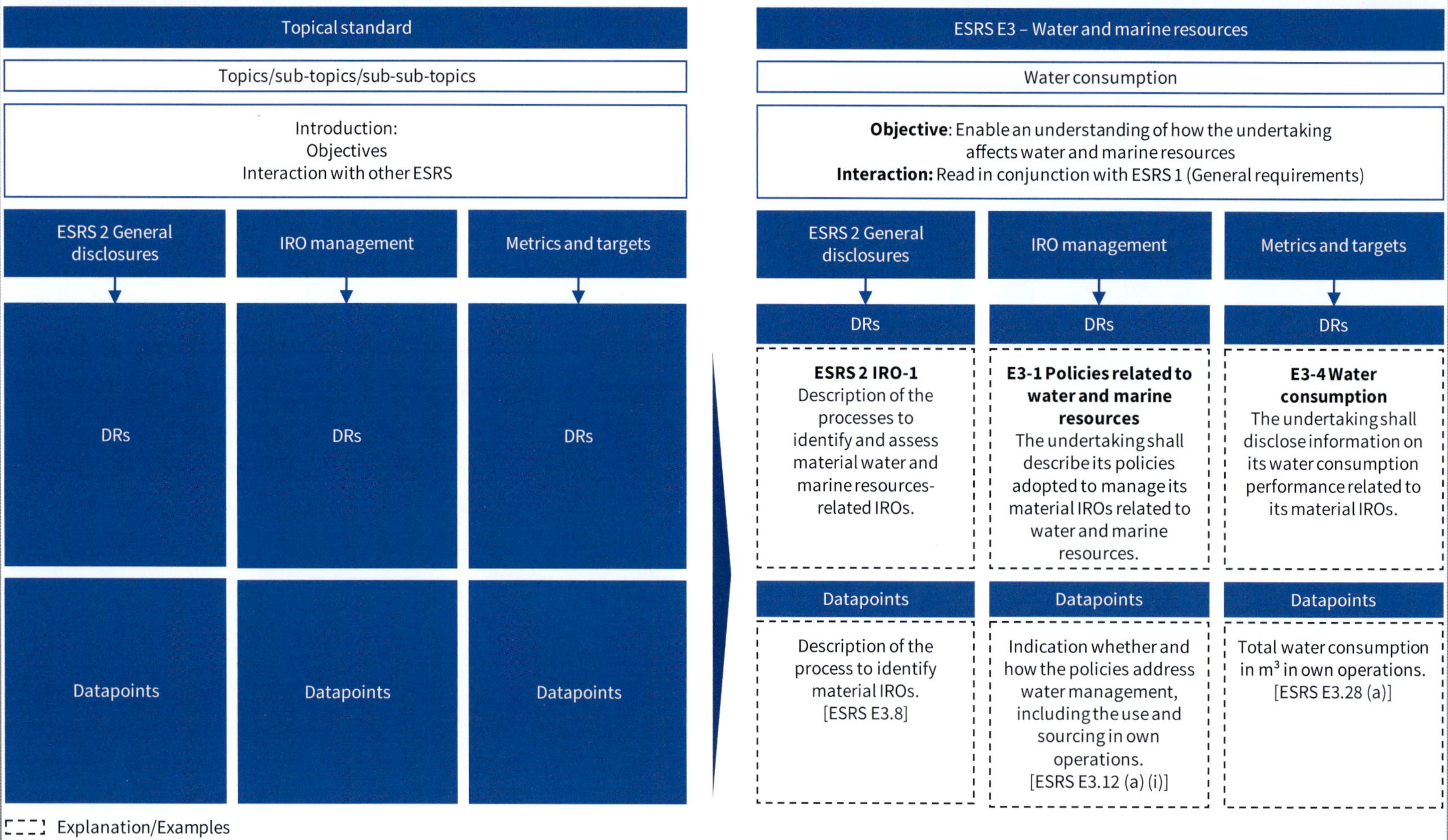

Intentionally left blank

Overview of ESRS (VIII/XIII)

Overview of sustainability matters

ESRS	Topic	Sub-topic	Sub-sub-topic
E1	Climate change	Climate change adaption	
		Climate change mitigation	
		Energy	
E2	Pollution	Pollution of air	
		Pollution of water	
		Pollution of soil	
		Pollution of living organisms and food resources	
		Substances of concern	
		Substances of very high concern	
		Microplastics	
E3	Water and marine resources	Water	Water consumption
			Water withdrawals
			Water discharges
			Water discharges in the oceans
		Marine resources	Extraction and use of marine resources

ESRS	Topic	Sub-topic	Sub-sub-topic
E4	Biodiversity and ecosystems	Direct impact drivers of biodiversity loss	Climate change
			Land-use change, fresh water-use change and sea-use change
			Direct exploitation
			Invasive alien species
			Pollution
			Others
		Impacts on the state of species	Species population size
			Species global extinction risk
		Impacts on the extent and condition of ecosystems	Land degradation
			Desertification
			Soil sealing
		Impacts and dependencies on ecosystem services	

[- - -] Explanation/Examples

Overview of ESRS (IX/XIII)

Overview of sustainability matters

ESRS	Topic	Sub-topic	Sub-sub-topic
E5	Circular economy	Resource inflows, including resource use	
		Resource outflows related to products and services	
		Waste	
S1	Own workforce	Working conditions	Secure employment
			Working time
			Adequate wages
			Social dialogue
			Freedom of association, the existence of work councils and the information, consultation and participation rights of workers
			Collective bargaining, including the rate of workers covered by collective agreements
			Work-life balance
			Health and safety

ESRS	Topic	Sub-topic	Sub-sub-topic
S1	Own workforce	Equal treatment and opportunities for all	Gender equality and equal pay for work of equal value
			Training and skills development
			Employment and inclusion of persons with disabilities
			Measures against violence and harassment in the workplace
			Diversity
		Other work-related rights	Child labour
			Forced labour
			Adequate housing
			Privacy

Overview of ESRS (X/XIII)

Overview of sustainability matters

ESRS	Topic	Sub-topic	Sub-sub-topic
S2	Workers in the value chain	Working conditions	Secure employment
			Working time
			Adequate wages
			Social dialogue
			Freedom of association, the existence of work councils and the information, consultation and participation rights of workers
			Collective bargaining
			Work-life balance
			Health and safety
		Equal treatment and opportunities for all	Gender equality and equal pay for work of equal value
			Training and skills development
			The employment and inclusion of persons with disabilities
S2	Workers in the value chain	Equal treatment and opportunities for all	Measures against violence and harassment in the workplace
			Diversity
		Other work-related rights	Child labour
			Forced labour
			Adequate housing
			Privacy
S3	Affected communities	Communities' economic, social and cultural rights	Adequate housing
			Adequate food
			Water and sanitation
			Land-related impacts
			Security-related impacts

Overview of ESRS (XI/XIII)

Overview of sustainability matters

ESRS	Topic	Sub-topic	Sub-sub-topic
S3	Affected communities	Communities' civil and political rights	Freedom of expression
			Freedom of assembly
			Impacts on human rights defenders
		Rights of indigenous peoples	Free, prior and informed consent
			Self-determination
			Cultural rights
S4	Consumers and end-users	Information-related impacts for consumers and/or end-users	Privacy
			Freedom of expression
			Access to (quality) information
		Personal safety of consumers and/or end-users	Health and safety
			Security of a person
			Protection of children
		Social inclusion of consumers and/or end-users	Non-discrimination
			Access to products and services
			Responsible marketing practices

ESRS	Topic	Sub-topic	Sub-sub-topic
G1	Business conduct	Corporate culture	
		Protection of whistle-blowers	
		Animal welfare	
		Political engagement	
		Management of relationships with suppliers including payment practices	
		Corruption and bribery	Prevention and detection including training
			Incidents

Overview of ESRS (XII/XIII)

Overview of disclosures

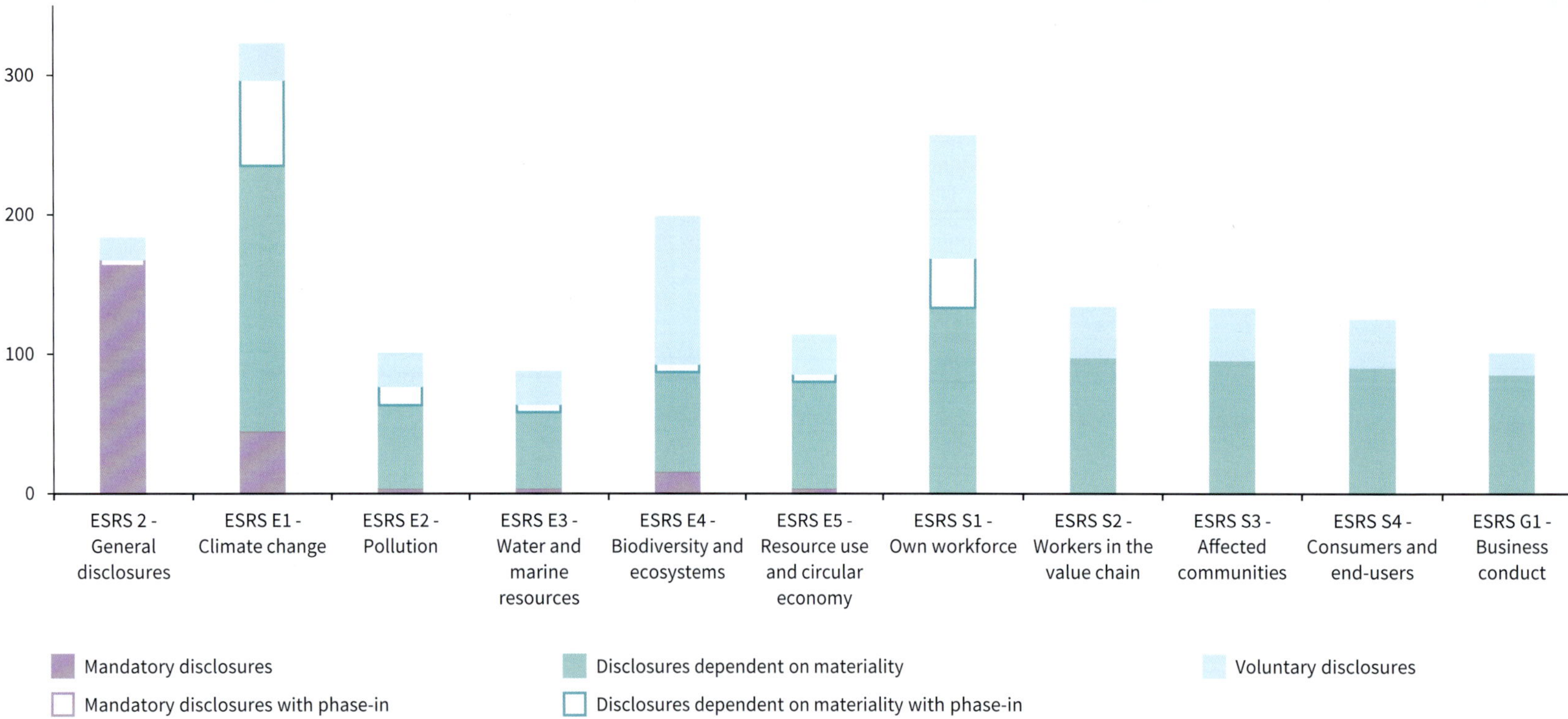

Note:
The number of disclosures was determined using professional judgement and depends on various factors, such as the level of detail and allocation methods. Due to the subjectivity involved, other counts may deviate.

ESRS 1 XXVIII/ XXIX

Note:
Certain standards (ESRS E1, E4, S1-S4) include a phase-in of one or two years for undertakings not exceeding the average number of 750 employees. In this diagram, when counting the disclosures and when presenting individual disclosures, these disclosures were not marked as "phased-in", as the phase-in does not apply to all undertakings. A detailed list of those phase-ins is included in ESRS 1 Appendix C.

Overview of ESRS (XIII/XIII)

Overview of key disclosures

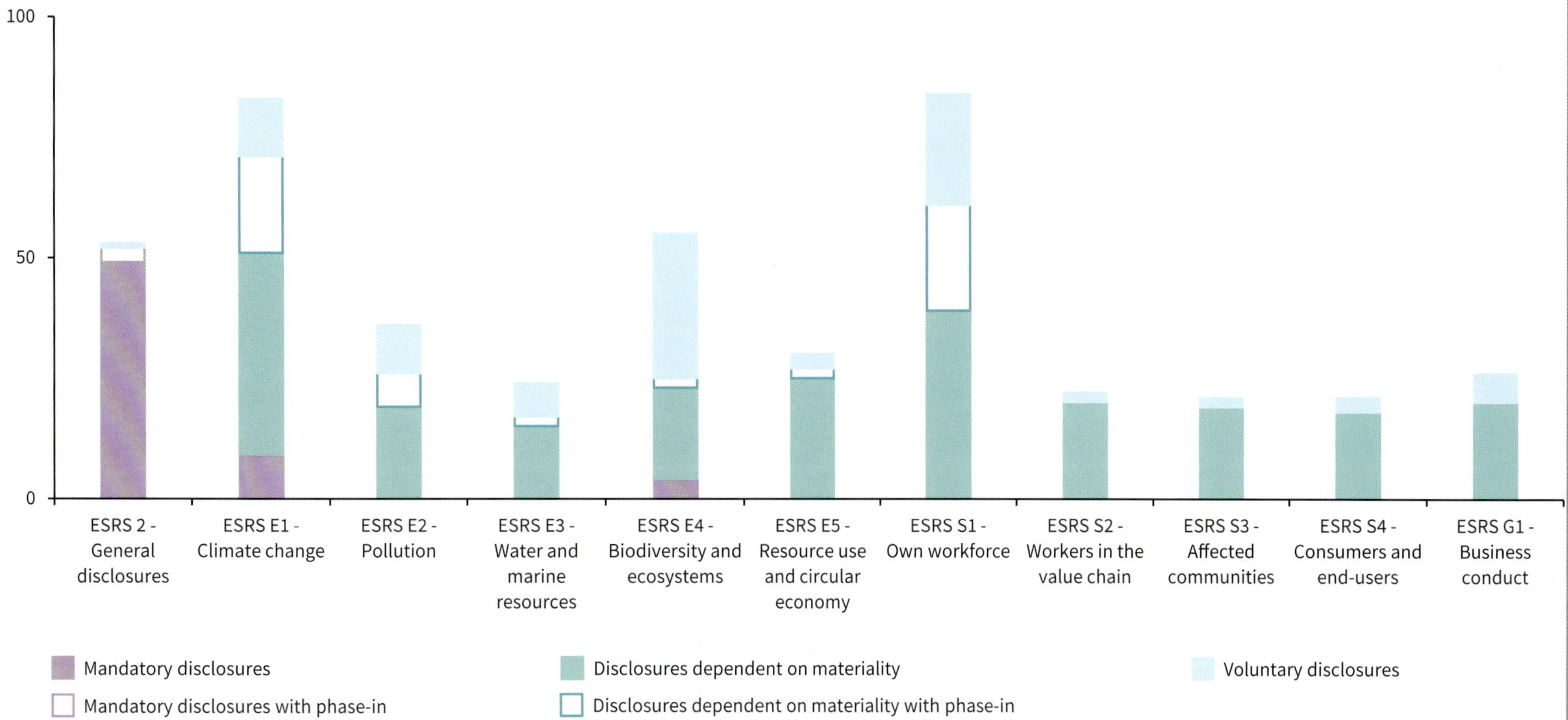

General information & structure I/II

Note:
The number of key disclosures was determined using professional judgement and depends on various factors, such as the level of detail and allocation methods. Due to the subjectivity involved, other counts may deviate.

ESRS 1 XXVIII/XXIX

Note:
Certain standards (ESRS E1, E4, S1-S4) include a phase-in of one or two years for undertakings not exceeding the average number of 750 employees. In this diagram, when counting the disclosures and when presenting individual disclosures, these disclosures were not marked as "phased-in", as the phase-in does not apply to all undertakings. A detailed list of those phase-ins is included in ESRS 1 Appendix C.

3 General information on ESRS – a visual approach

General information and structure of ESRS – a visual approach (I/II)

General information on ESRS – a visual approach

General information

Completeness	This publication does not contain all DRs resulting from the ESRS. Therefore, not all information of the ESRS Standards is included. The key disclosures include all datapoints that fulfil the definition of a key disclosure. Details on DRs explain more complex DRs by presenting additional disclosures and information as well as examples beyond the key disclosures.
Wording	"Among others", "in particular" or "e.g." indicate that not all items were included.
Units	Physical units (kg, tons, etc.) are only included in details on DRs.
Sources	The content of this document refers to the ESRS Standards in the versions listed in the chapter "Publication dates of the standards". Other considerations may apply if the listed standards are amended or additional standards are published. The use of other sources will be specifically mentioned, as applicable.
Tools and databases	The ESRS Standards mention tools and databases that provide access to more granular, specialised datasets that can further add additional information to disclosures. The most relevant references are included in the section "References to tools and databases".
Key disclosures	Within ESRS – a visual approach key disclosures are selected disclosures taken from the ESRS Standards that can be used to measure and manage the sustainability performance of an undertaking. Following this principle, they are likely to represent the most essential disclosures of a standard but are not intended to be a complete list of disclosures. Accompanying disclosures are not included in this definition.
Overall evaluation	The overall evaluation contains a subjective assessment of the corresponding standard as narrative or quantitative and a count of all disclosures included in the corresponding standard. The number of disclosures was determined using professional judgement and depends on various factors. Due to the subjectivity involved, other counts may deviate.

General information and structure of ESRS – a visual approach (II/II)

Structure of ESRS – a visual approach

Structure across standards	
Overview of sustainability matters of all E-/S-/G-Standards	Introduction to E-/S- and G-Standards by indicating the topics, sub-topics and sub-sub-topics (sustainability matters) of the corresponding standard, including references to the pages on the standards in this publication.
Cross-reference between ESRS 2 and all E-/S-/G-Standards	Overview of the reporting areas or chapters covered in the cross-cutting and E-/S- and G-Standards, including references to the pages on the respective DRs in this publication.

Structure of each cross-cutting and topical standard	
Introduction	Description of the objective, sustainability matters, conjunctions and intersectional topics covered by other standards, interactions with other frameworks, applicability (phase-ins that are not applicable to all undertakings) as well as an overall evaluation of the nature and content of the standard and a disclosure count. The number of disclosures was determined using professional judgement and depends on various factors. Due to the subjectivity involved, other counts may deviate.
Structure and content	Overview and brief explanation of all DRs of the standard, including references to pages with details on DRs and applicable DRs of ESRS 2. In some cases, the titles of the DRs have been editorially adapted to increase readability.
Key disclosures concerning the reporting areas	Disclosures of the cross-cutting and topical standards that can be used to measure and manage the sustainability performance of the undertaking. Accompanying disclosures are not included in this presentation.
Details on DRs	Presentations for complex DRs with additional disclosures beyond the key disclosures presented in order to understand the context of the DR.

Colour scheme of the ESRS – a visual approach (I/I)

Explanation of shapes and colours

Shapes and colours	Explanation
Disclosure	Any disclosures resulting from the ESRS cross-cutting and topical standards. Related disclosures have been aggregated in some cases.
Explanation/Examples	Further explanatory information or examples.
Breakdowns	Breakdowns are separate parts of a disclosure that add up to the total information (e.g. breakdown of energy consumption by sources).
Mandatory disclosure	Indicates that the disclosure is mandatory without any phase-in period. Mandatory disclosures shall be made irrespective of the materiality of the sustainability matter.
Mandatory disclosure with phase-in	Indicates that the disclosure is mandatory with a phase-in period. ESRS 1 XXVIII/ XXIX

Shapes and Colours	Explanation
Disclosure dependent on materiality	Indicates that the disclosure is dependent on materiality assessment without any phase-in period.
Disclosure dependent on materiality with phase-in	Indicates that the disclosure is dependent on materiality assessment with a phase-in period. ESRS 1 XXVIII/ XXIX
Voluntary disclosure	Indicates that the disclosure is voluntary.
1 2	Requirements that may, if applicable and in combination with other DRs, lead to (1) disclosures dependent on materiality or (2) voluntary disclosures based on ESRS 1.

Note:
Certain standards (ESRS E1, E4, S1-S4) include a phase-in of one or two years for undertakings not exceeding the average number of 750 employees. [ESRS 1 Appendix C] When counting the disclosures and when presenting individual disclosures, these disclosures were not marked as "phased-in", as the phase-in does not apply to all undertakings. The respective standards contain an applicability note that refers to this fact.

Referencing (I/I)

Explanation of references

Exemplary reference	Explanation
[E1-1]	Reference to a DR
[65]	Reference to a text passage within the same standard
[ESRS 2.63]	Reference to a text passage within another standard
[AR 19]	Reference to an AR within the same standard
[1.2]	Reference to a chapter within the same standard
[ESRS 2 MDR-P]	Reference to a chapter within another standard
DR E1-6 – 1/2	Indicates that the presentation of the DR contains two pages

Exemplary reference	Explanation
IV/VI	Reference to a page number within the same standard
E1-I/VI	Reference to another standard and page number of the referenced standard
ESRS 2 MDR-P	Reference to a chapter within another standard
ESRS 2 MDR-P	Reference to the Minimum Disclosure Requirements (MDR) for policies, actions, metrics and targets that refer to each material sustainability matter in each ESRS and to entity-specific disclosures. In some cases, the DR of the ESRS cannot be assigned as a whole and clearly to one of these areas. In these cases, they were assigned to the area to which they predominantly correspond. An individual assessment at the level of the datapoints to identify the relevant MDR may be necessary

References indicate where additional information is available

4 Cross-cutting

ESRS 1 – ESRS 2 (I/I)

ESRS 1 – General requirements

ESRS 1
I/XXIX

1.	2.	3.	4.	5.	6.	7.	8.	9.	10.
Categories of ESRS Standards, reporting areas and drafting conventions	Qualitative characteristics of information	Double materiality as the basis for sustainability disclosures	Due diligence	Value chain	Time horizons	Preparation and presentation of sustainability information	Structure of the sustainability statement	Linkages with other parts of corporate reporting and connected information	Transitional provisions

General requirements for the preparation and presentation of sustainability statements that shall be applied across all ESRS

ESRS 2 – General disclosures

ESRS 2
I/XXII

1. Basis for preparation		2. Governance					3. Strategy			4. Impact, risk and opportunity management				5. Metrics and targets	
BP-1	BP-2	GOV-1	GOV-2	GOV-3	GOV-4	GOV-5	SBM-1	SBM-2	SBM-3	IRO-1	IRO-2	MDR-P	MDR-A	MDR-M	MDR-T

DRs that refer to the undertaking as a whole (BP-1 to IRO-2)

MDR for policies, actions, metrics and targets that refer to each material sustainability matter in each ESRS and to entity-specific disclosures (MDR-P to MDR-T)

Note:
Both ESRS 1 and ESRS 2 are referred as cross-cutting standards, meaning that they apply to all sustainability matters and should be read in conjunction with all topical standards.

☐ DR

ESRS 1 – General requirements

Objective:

Enabling understanding of:

- The architecture of ESRS, the drafting conventions and fundamental concepts used and the general requirements for preparing and presenting sustainability information in accordance with the Accounting Directive.
- The ESRS reporting areas, engagement with stakeholders, the concept of double materiality, how to report on value chain information, which time horizons should be used for reported information, how to use estimates and how the sustainability statement should be structured.

Sustainability matters: All

Interaction with other ESRS:

- **Conjunctions**: The general requirements of ESRS 1 shall be applied to all topical and sector-specific ESRS as well as to entity-specific disclosures.
- **Intersectional topics covered by other standards**: n/a

Interaction with other frameworks:

- Accounting Directive
- Council Decision on the security rules for protecting EU classified information
- EMAS
- EU Taxonomy
- GRI Sector Standards
- GRI Standards
- ISSB
- OECD Guidelines for Multinational Enterprises
- Regulation establishing the European Defence Fund
- Sustainable Finance Disclosure Regulation (SFDR)
- UNGPs on Business and Human Rights

Overall evaluation:

Narrative **Quantitative**

Requirements that may, if applicable and in combination with other DRs, lead to a disclosure obligation subject to materiality

Requirements that may, if applicable and in combination with other DRs, lead to a voluntary disclosure

The overall evaluation contains a subjective assessment of the corresponding standard as narrative or quantitative.

ESRS 1 – General requirements (ESRS 1 – I/XXIX)

Structure and content

Chapter 1 (II/XXIX)

Categories of ESRS Standards, reporting areas and drafting conventions

Introduction of the categories of ESRS Standards (cross-cutting standards, topical standards, sector-specific standards, entity-specific disclosures), the reporting areas of the sustainability reporting including MDRs and drafting conventions.

Chapter 2 (VI/XXIX)

Qualitative characteristics of information

The sustainability statement shall be based on qualitative characteristics of information that are fundamental, such as relevance and faithful representation, and enhancing, such as comparability, verifiability and understandability.

Chapter 3 (VII/XXIX)

Double materiality as the basis for sustainability disclosures

Sustainability matters shall be assessed based on impact materiality and financial materiality (double materiality concept). This chapter also provides information about stakeholders and their relevance to the materiality assessment process as well as the level of disaggregation.

Chapter 4 (XIV/XXIX)

Due diligence

Information on the undertaking's due diligence process including the process to identify, prevent, mitigate and account for how the undertaking addresses its negative impacts on the environment and people along with risk and opportunities from its economic activities.

Chapter 5 (XV/XXIX)

Value chain

Material IROs connected with the undertaking's value chain(s) shall be included in the reporting undertaking's sustainability statement. If the undertaking cannot collect value chain information, under certain circumstances it may use estimates.

Chapter 6 (XVII/XXIX)

Time horizons

The reporting period for the sustainability and annual financial statement shall be consistent, links shall be established between retrospective and forward-looking information, progress is to be benchmarked against a base year and specific definitions of time horizons shall be adopted.

Chapter 7 (XVIII/XXIX)

Preparation and presentation of sustainability information

The undertaking shall disclose comparative information in respect of the previous period and the sources of estimates and outcome certainty. Prior period errors shall be corrected and the materiality assessment shall be made at group level, if applicable. The undertaking may omit classified, sensitive and information corresponding to intellectual property.

Chapter 8 (XXIII/XXIX)

Structure of the sustainability statement

The sustainability statement shall be reported in a dedicated section of the management report, including disclosures as per the EU Taxonomy, and it should be structured into four sections: general, environmental, social and governance information.

Chapter 9 (XXIV/XXIX)

Linkages with other parts of corporate reporting and connected information

Links between sustainability information and related information in corporate reporting shall be ensured by reference, and connected information shall be presented narratively, e.g. by linking information on governance, strategy and risk management to related metrics and targets.

Chapter 10 (XXVI/XXIX)

Transitional provisions

Undertakings may use transitional provisions for entity-specific disclosures, value chain information, reporting of comparative information and a list of phased-in DRs.

ESRS 1 – General requirements (ESRS 1 – II/XXIX)

Categories of ESRS Standards, reporting areas and drafting conventions [1.]

Categories of ESRS Standards [1.1]

Category	Explanation/Examples
ESRS 1 – General requirements	Describes the architecture of ESRS Standards, explains drafting conventions and fundamental concepts and sets out general requirements for preparing and presenting sustainability-related information.
ESRS 2 – General disclosures	Establishes DRs that shall be provided at a general level across all material sustainability matters on the reporting areas governance, strategy, IRO management and metrics and targets.
Topical ESRS	Cover E-/S- and G-Standards and can additionally include specific requirements that complement the general level DRs of ESRS 2.
Sector-specific ESRS	Address undertakings operating in specific sectors and address IROs that are likely to be material for all undertakings in a specific sector and that are not (sufficiently) covered by topical ESRS. Sector-specific ESRS were not yet published at the date of this publication.
Entity-specific disclosures	If a material IRO is not covered or not covered with sufficient granularity by an ESRS, the undertaking shall provide additional entity-specific disclosures to enable users to understand the undertaking's IROs by disclosing, where applicable, all material information on the reporting areas governance, strategy, IRO management and metrics and targets. (V/XXIX)

Explanation/Examples

ESRS 1 – General requirements (ESRS 1 – III/XXIX)

Categories of ESRS Standards, reporting areas and drafting conventions [1.]

Reporting areas [1.2]

Reporting areas relevant for the structure of every ESRS

Governance (GOV)	Governance processes, controls and procedures used to monitor, manage and oversee IROs.
Strategy (SBM)	Interactions of strategy and business model with material IROs, including respective management strategies.
IRO management (IRO)	Processes to identify IROs and to assess their materiality.
	Processes to manage material sustainability matters through policies and actions.
Metrics and targets (MT)	The undertaking's performance, including targets it has set and progress towards meeting them.

MDRs

ESRS 2 I/XXII

MDRs on Policies (MDR-P)	ESRS 2 includes specific MDRs. The undertaking shall apply those regarding policies, actions, metrics and targets together with the corresponding DRs in topical and sector-specific ESRS.
MDRs on Actions (MDR-A)	
MDRs on Targets (MDR-T)	
MDRs on Metrics (MDR-M)	

[- -] Explanation/Examples

ESRS 1 – General requirements (ESRS 1 – IV/XXIX)

Categories of ESRS Standards, reporting areas and drafting conventions [1.]

Drafting conventions [1.3]

Definition of IROs

Impacts	Actual or potential positive or negative short-, medium- or long-term effects of the undertaking on the environment and people connected with the undertaking's own operations and upstream and downstream value chain, through its products and services, as well as through its business relationships. The impacts can be intended or unintended and reversible or irreversible. [Annex II]
Risks	Sustainability-related risks with negative financial effects arising from environmental, social or governance matters that may negatively affect the undertaking's financial position, financial performance, cash flows, access to finance or cost of capital over the short-, medium- or long-term. [Annex II]
Opportunities	Sustainability-related opportunities with positive financial effects arising from environmental, social or governance matters that may positively affect the undertaking's financial position, financial performance, cash flows, access to finance or cost of capital over the short-, medium- or long-term. [Annex II]

Degree of obligations used in the DRs

Wording used in the ESRS	Resulting degree of obligation	Reference to the colour scheme (Colour scheme I/I)
Shall disclose	Provision is prescribed by a DR or datapoint.	Mandatory disclosure / Mandatory disclosure with phase-in / Disclosure dependent on materiality / Disclosure dependent on materiality with phase-in
May disclose	Voluntary disclosure to encourage good practice.	Voluntary disclosure
Shall consider	Issues, resources or methodologies that the undertaking is expected to take into account or use in the preparation of disclosures.	Explanation/Examples

Legend: Mandatory disclosure | Mandatory disclosure with phase-in | Disclosure dependent on materiality | Disclosure dependent on materiality with phase-in | Voluntary disclosure | Explanation/Examples

ESRS 1 – General requirements (ESRS 1 – V/XXIX)

Entity-specific disclosures

In addition to the disclosures laid down in the ESRS, when an undertaking concludes that an IRO is not covered or not covered with sufficient granularity by an ESRS but is material due to its specific facts and circumstances, it shall provide additional entity-specific disclosures to enable users to understand the undertaking's sustainability-related IROs [11]

All material IROs

Material IROs sufficiently covered by topical ESRS	IROs not covered or not sufficiently covered by topical ESRS
Disclosures related to ESRS E1-E5, ESRS S1-S4, ESRS G1	Entity-specific disclosures

Objective

The entity-specific disclosures shall enable users to understand the undertaking's IROs in relation to environmental, social or governance matters. [AR 1]

The extent to which sustainability matters are covered by ESRS is expected to evolve as further DRs are developed. Therefore, the need for entity-specific disclosures is likely to decrease over time, in particular as a result of the future adoption of sector specific standards. [130]

General remarks

VI/XXIX

The undertaking shall ensure that:

- The disclosures meet the qualitative characteristics of information as set out in [2., Appendix B]
- Where applicable, the disclosures include all material information related to the reporting areas of governance, strategy, IRO management and metrics and targets.
- Further guidance for developing entity-specific disclosures can be found by considering the information required under topical ESRS that addresses similar sustainability matters. [AR 5]

Entity-specific metrics

When disclosing entity-specific metrics the undertaking shall consider: [AR 3]

- For **impacts**: Whether its chosen performance metrics provide insight into how effective the undertaking's practices are in reducing negative outcomes and/or increasing positive outcomes for people and the environment.
- For **risks and opportunities**: The likelihood that the practices result in financial effects.
- Whether the measured outcomes are sufficiently reliable (no excessive number of assumptions and unknowns that would render the metrics too arbitrary to provide a faithful representation).
- Whether sufficient contextual information is provided to interpret performance metrics appropriately and whether variations in such contextual information may impact the comparability of the metrics over time.

Comparability

The undertaking shall carefully consider:

- Comparability between undertakings, while still ensuring relevance of the information provided, recognising that comparability may be limited for entity-specific disclosures.
- Whether the available and relevant frameworks, initiatives, reporting standards and benchmarks (such as technical material issued by the ISSB or the Global Report Initiative (GRI)) provide elements that can support comparability to the maximum extent possible.
- Comparability over time: Consistency of methodologies and disclosures is a key factor for achieving comparability over time.

XXVI/XXIX

Note:
Entity-specific disclosures shall be disclosed alongside the most relevant sector-agnostic and sector-specific disclosures. [117] Transitional provisions concerning entity-specific disclosures may be considered.

● Requirements that may, if applicable and in combination with other DRs, lead to disclosures dependent on materiality

[- -] Explanation/Examples

ESRS 1 – General requirements (ESRS 1 – VI/XXIX)

Qualitative characteristics of information [2., Appendix B]

When preparing its sustainability statement, the undertaking shall apply fundamental and enhancing qualitative characteristics of information

Fundamental qualitative characteristics	Relevance	Sustainability information that (1) may make a **difference** in the decisions of users under a double materiality approach, (2) has **predictive** value, **confirmatory** value or both, (3) **confirms or changes** previous assessments or (4) is assessed to be material due to the nature or magnitude or both of **entity-specific aspects** for the undertaking. [QC 1-4]
	Faithful representation	Sustainability information that is, among others, (1) **complete** by including all material information necessary to understand that IRO, including the adaption of strategy, risk management and governance in relation to IROs, as well as metrics identified to set targets and measure performance, (2) **neutral** by following a principle of prudence and without being biased, and (3) **accurate** by disclosing accurate information gathered through adequate processes and internal controls to avoid material errors or misstatements. The undertaking may present net information, in addition to gross values, if such presentation **does not obscure relevant information**. [QC 5-9]
Enhancing qualitative characteristics	Comparability	**Comparability** of sustainability information with preceding **periods** or with **peers**. Consistency in methods helps to achieve comparability. Nevertheless, comparability does not mean uniformity. [QC 10-12]
	Verifiability	Sustainability information is **verifiable** if it is possible to corroborate the information itself or the inputs used to derive it. Verifiability means that various knowledgeable and independent observers **could reach consensus**, although not necessarily complete agreement, that a particular depiction is a faithful representation. Among others, this can be achieved by including information that can be **corroborated** by comparing it with other information available to users and by the disclosure of inputs and methods of calculation used to produce **estimates** or approximations. [QC 13-15]
	Understandability	**Understandable** information is characterised by (1) **clarity** by refraining from generic phrases, duplications and ambiguous language and (2) **conciseness** by reporting only material information. Sustainability information shall be disclosed as a coherent whole by, among others, contextualising reported information and, if applicable, presenting links to the financial statements. [QC 16-20]

Explanation/Examples

ESRS 1 – General requirements (ESRS 1 – VII/XXIX)

Double materiality as the basis for sustainability disclosures [3.]

Stakeholders and their relevance to the materiality assessment process [3.1]

Stakeholder groups

Affected stakeholders

Whose interests are or could be **positively or negatively affected** by the undertaking's activities and its direct and indirect business relationships across its value chain. Nature may be considered as a silent stakeholder. In this case, ecological data and data on the conservation of species may support the undertaking's materiality assessment. [AR 7]

Users of sustainability statements

Primary users of general-purpose financial reporting (among others existing and potential investors) and **other users** of sustainability statements including, among others, the undertaking's business partners, trade unions, social partners, civil society and non-governmental organisations (NGO).

Engagement with affected stakeholders

The importance of stakeholder engagement

The engagement and the assessment of the affected stakeholder's views and interests is **central** to the due diligence process and materiality assessment. This includes processes to identify and assess actual and potential negative impacts which then inform the assessment process to identify the material impacts for the purposes of sustainability reporting.

Dialogue with affected stakeholders

Materiality assessment is **informed** by the dialogue with affected stakeholders. The undertaking may engage with affected stakeholders or their representatives (such as employees or trade unions), along with users of sustainability reporting and other experts, to provide inputs or feedback on its conclusions regarding its material IROs.

Note:
ESRS do not require to actively engage in dialogue with affected stakeholders. There are no specific behaviour requirements on stakeholder engagement or the wider due diligence process. Hence, the undertaking is also allowed to consider interests and views of stakeholders without actively communicating with them. However, the Accounting Directive requires the management of the undertaking to inform the workers' representatives at the appropriate level and discuss with them the relevant information and the means of obtaining and verifying sustainability information. [Articles 19a paragraph 5, 29a paragraph 5 Accounting Directive].

Explanation/Examples (dashed boxes)

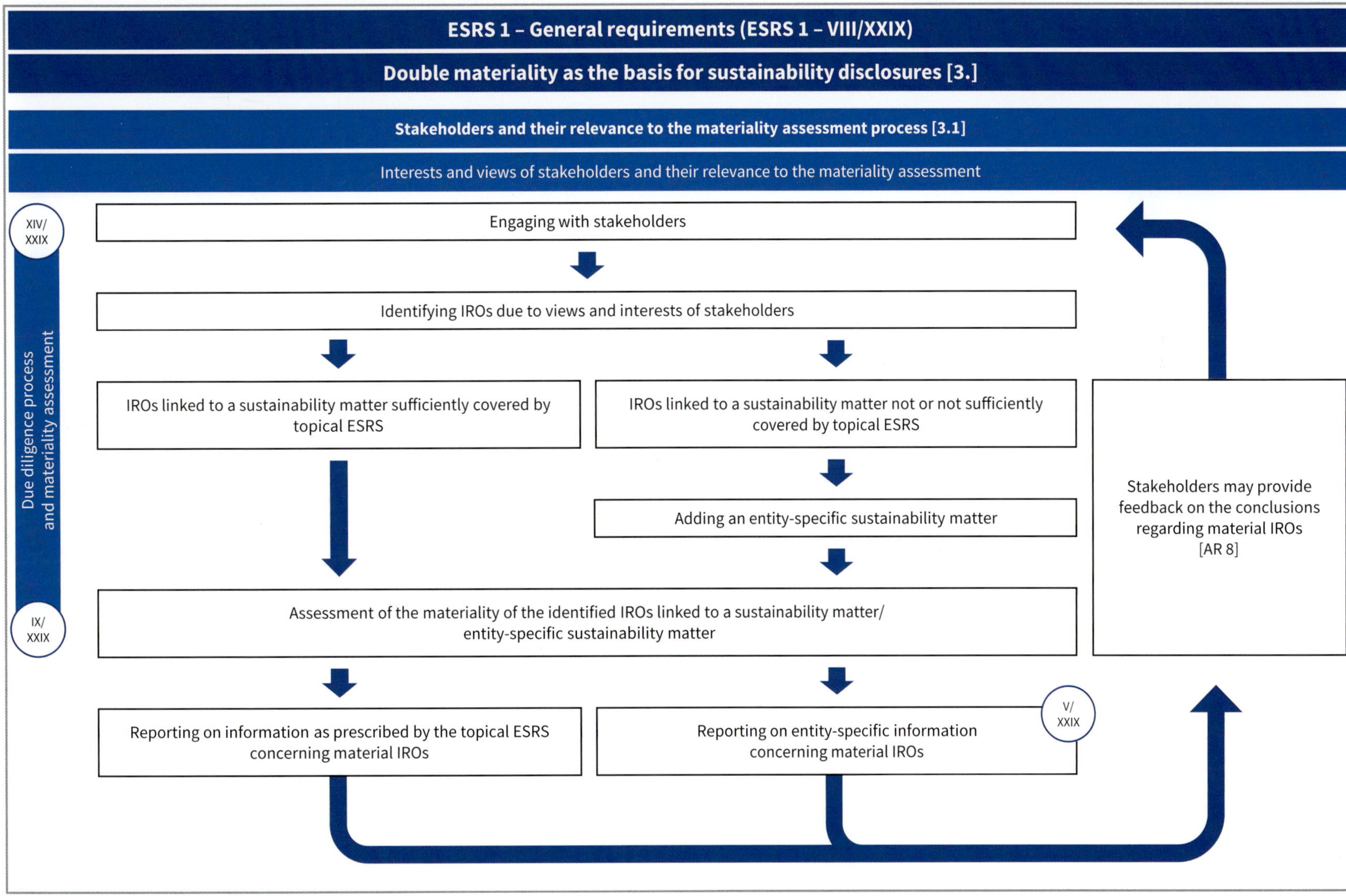
ESRS 1 – General requirements (ESRS 1 – VIII/XXIX)
Double materiality as the basis for sustainability disclosures [3.]
Stakeholders and their relevance to the materiality assessment process [3.1]
Interests and views of stakeholders and their relevance to the materiality assessment
XIV/ XXIX
Due diligence process and materiality assessment
IX/ XXIX
Engaging with stakeholders
Identifying IROs due to views and interests of stakeholders
IROs linked to a sustainability matter sufficiently covered by topical ESRS
IROs linked to a sustainability matter not or not sufficiently covered by topical ESRS
Adding an entity-specific sustainability matter
Assessment of the materiality of the identified IROs linked to a sustainability matter/ entity-specific sustainability matter
Reporting on information as prescribed by the topical ESRS concerning material IROs
Reporting on entity-specific information concerning material IROs
V/ XXIX
Stakeholders may provide feedback on the conclusions regarding material IROs [AR 8]

ESRS 1 – General requirements (ESRS 1 – IX/XXIX)

Double materiality as the basis for sustainability disclosures [3.]

A sustainability matter is material if it is linked to material impacts and/or material risks and opportunities (double materiality).

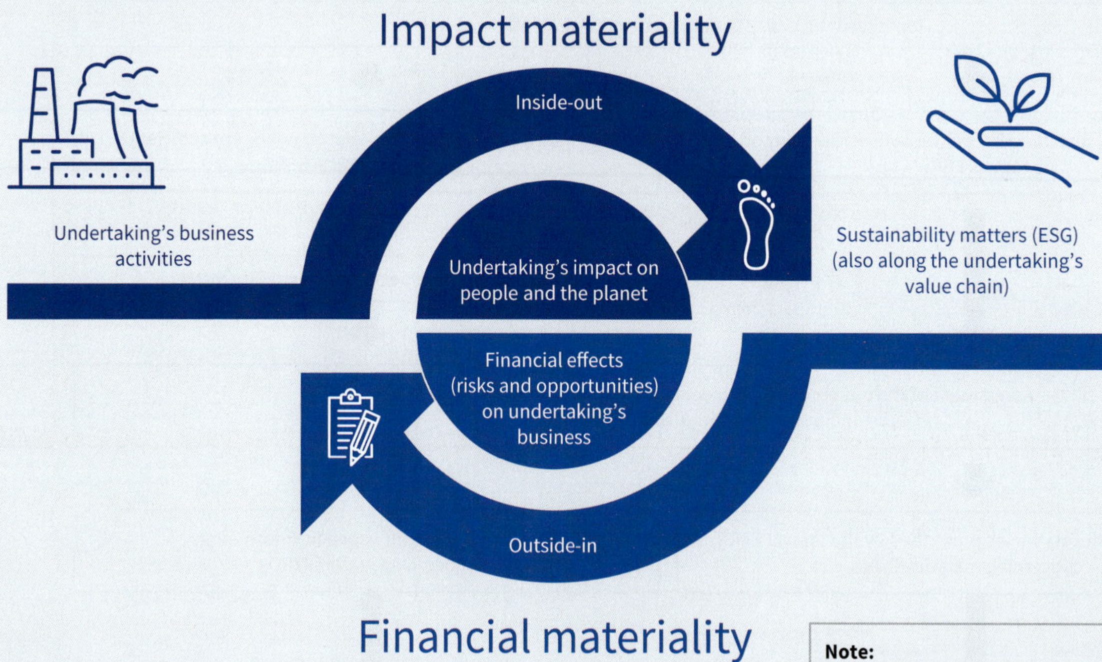

Note:
An undertaking's impact on people or the environment can in turn result in material financial effects on the undertaking.

ESRS 1 – General requirements (ESRS 1 – X/XXIX)

Double materiality as the basis for sustainability disclosures [3.]

Impact materiality [3.4]

A sustainability matter is material from an impact perspective when it pertains to the undertaking's material actual or potential, positive or negative impacts on people or the environment over the short-, medium- or long-term

Potential impacts:

Likelihood of the impact [45]

&

Actual impacts:

Severity of the impact [45, 46]

Negative impacts:

- **Scale:** How grave the negative impact is for people or the environment [AR 10 (a)]
- &
- **Scope:** How widespread the negative impacts are [AR 10 (b)]
- &
- **Irremediable character:** Whether and to what extent the negative impacts could be remediated [AR 11 (c)]

Positive impacts:

- **Scale:** How beneficial the positive impact is for people or the environment [AR 10 (a)]
- &
- **Scope:** How widespread the positive impacts are [AR 10 (b)]

Financial materiality [3.5]

A sustainability matter is material from a financial perspective if it triggers or could reasonably be expected to trigger material financial effects on the undertaking over the short-, medium- or long-term

Likelihood of occurrence of the financial effect [51]

&

Potential **magnitude** of the financial effects [51]

Financial effects:

Risks or opportunities that have (or could reasonably be expected to have) a **material influence** on the undertaking's:

Development or **Financial position** or **Cash flows** or **Access to finance** or **Cost of capital**

Over the **short-, medium- or long-term**

Sources of financial effects:

- **Dependence on resource availability** (including quality and pricing) [AR 13 (a)]
- **Changes in regulations**, when the undertaking's activities result in negative impacts [AR 13 (b)]
- **Dependence on supplier relationships** (including acceptable terms) [AR 13 (c)]

Note:
Any of the three characteristics (scale, scope and irremediable character) can make a negative impact severe. In the case of a potential negative human rights impact, the severity of the impact takes precedence over its likelihood. [AR 11]

[dashed box] Explanation/Examples

ESRS 1 – General requirements (ESRS 1 – XI/XXIX)

Double materiality as the basis for sustainability disclosures [3.]

Material impacts or risks arising from actions to address sustainability matters [3.6]

The materiality assessment process may lead to the identification of situations in which actions to address impacts or risks, or to benefit from certain opportunities in relation to a sustainability matter, might have material negative impacts or cause material risks in relation to sustainability matters

Case examples

An action plan to decarbonise production that involves abandoning certain products might have material negative impacts on the undertaking's own workforce and results in material risks due to redundancy payments.

An action plan of an automotive supplier to focus on the supply of e-vehicles might lead to stranded assets for the production of supply parts for conventional vehicles.

Disclosure obligation

In such situations, the undertaking shall:

- Disclose the existence of material negative impacts or material risks together with the actions that generate them, with a cross-reference to the topic to which the impacts or risks relate and
- Provide a description of how the material negative impacts or material risks are addressed under the topic to which they relate.

● Requirements that may, if applicable and in combination with other DRs, lead to disclosures dependent on materiality

[- -] Explanation/Examples

ESRS 1 – General requirements (ESRS 1 – XII/XXIX)

Double materiality as the basis for sustainability disclosures [3.]

Level of disaggregation [3.7]

When needed for a proper understanding of the material IROs, the undertaking shall disaggregate the reported information

Specificity and context	The undertaking shall ensure that the level of aggregation does not obscure the specificity and context necessary to interpret the information.
	The undertaking shall not aggregate material items that differ in nature.
Level of disaggregation	Disaggregation by • Country when there are significant variations of material IROs or • Significant site/asset when material IROs are highly dependent to a specific location/asset or • Subsidiary, depending on facts and circumstances The level and nature of disaggregation adopted in the materiality assessment shall be considered.
	When a topical/sector-specific ESRS requires that a specific level of disaggregation is adopted in preparing a specific item of information, this requirement shall prevail.

Note:
Example concerning the level of disaggregation adopted in the materiality assessment: An undertaking in the clothing industry operates in different countries where decent working conditions and reasonable wages are ensured. However, if there is a location, e.g. in a country where the undertaking's subsidiary pays less than the country's adequate wages, the materiality assessment should be performed at this level of disaggregation to identify the material negative impacts.

Explanation/Examples

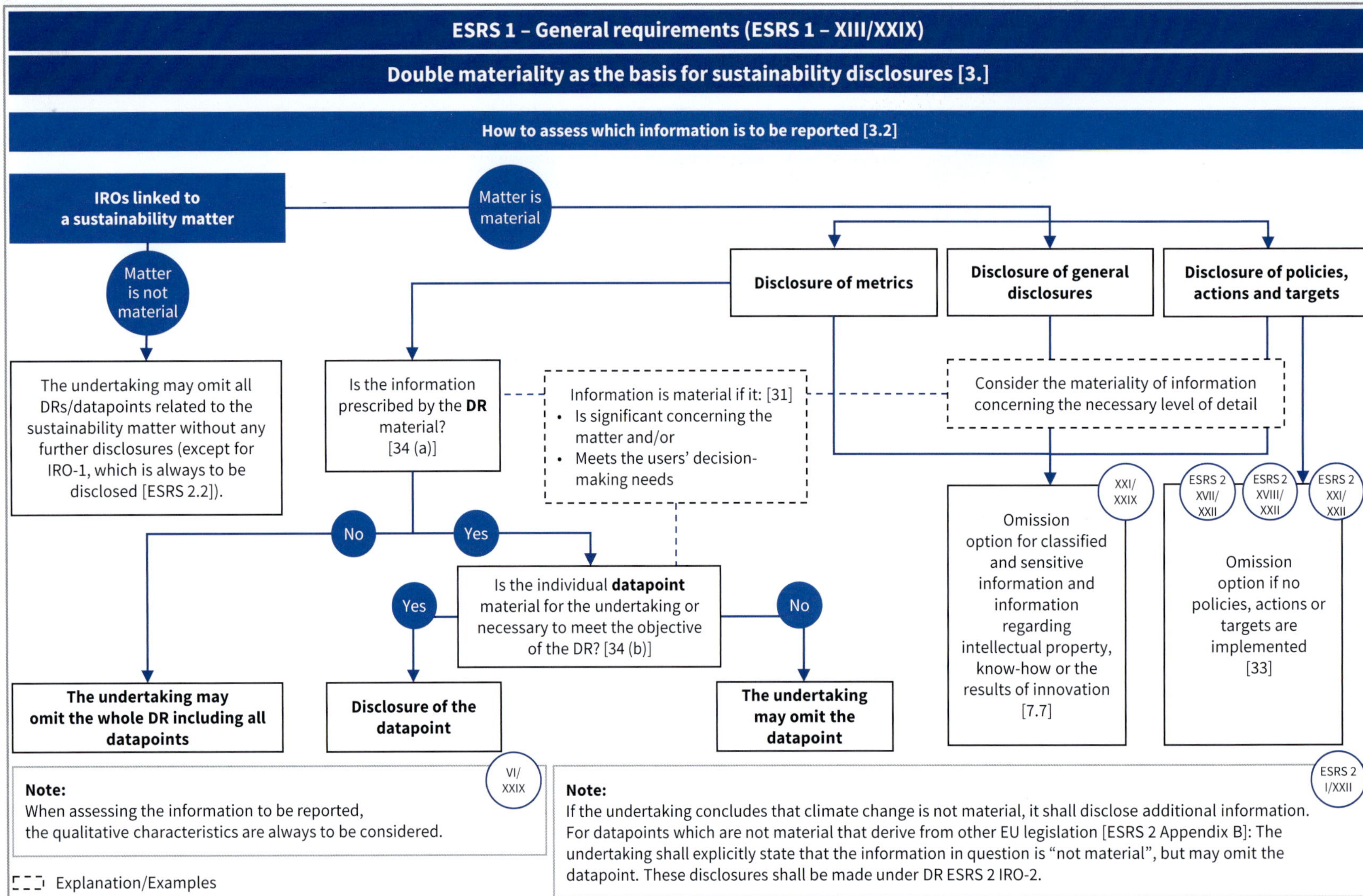
ESRS 1 – General requirements (ESRS 1 – XIII/XXIX)
Double materiality as the basis for sustainability disclosures [3.]
How to assess which information is to be reported [3.2]
IROs linked to a sustainability matter
Matter is material
Matter is not material
The undertaking may omit all DRs/datapoints related to the sustainability matter without any further disclosures (except for IRO-1, which is always to be disclosed [ESRS 2.2]).
Disclosure of metrics
Disclosure of general disclosures
Disclosure of policies, actions and targets
Is the information prescribed by the DR material? [34 (a)]
Information is material if it: [31]
• Is significant concerning the matter and/or
• Meets the users' decision-making needs
Consider the materiality of information concerning the necessary level of detail
No
Yes
Is the individual datapoint material for the undertaking or necessary to meet the objective of the DR? [34 (b)]
Yes
No
The undertaking may omit the whole DR including all datapoints
Disclosure of the datapoint
The undertaking may omit the datapoint
XXI/ XXIX
Omission option for classified and sensitive information and information regarding intellectual property, know-how or the results of innovation [7.7]
ESRS 2 XVII/ XXII
ESRS 2 XVIII/ XXII
ESRS 2 XXI/ XXII
Omission option if no policies, actions or targets are implemented [33]
VI/ XXIX
Note:
When assessing the information to be reported, the qualitative characteristics are always to be considered.
ESRS 2 I/XXII
Note:
If the undertaking concludes that climate change is not material, it shall disclose additional information. For datapoints which are not material that derive from other EU legislation [ESRS 2 Appendix B]: The undertaking shall explicitly state that the information in question is "not material", but may omit the datapoint. These disclosures shall be made under DR ESRS 2 IRO-2.
Explanation/Examples

ESRS 1 – General requirements (ESRS 1 – XIV/XXIX)

Due diligence [4.]

Due diligence is a process by which the undertaking identifies, prevents, mitigates and accounts for how it addresses actual or potential negative impacts on people and on the environment connected with its business. This includes its own operations as well as the upstream and downstream value chain.

Presentation of the due diligence process

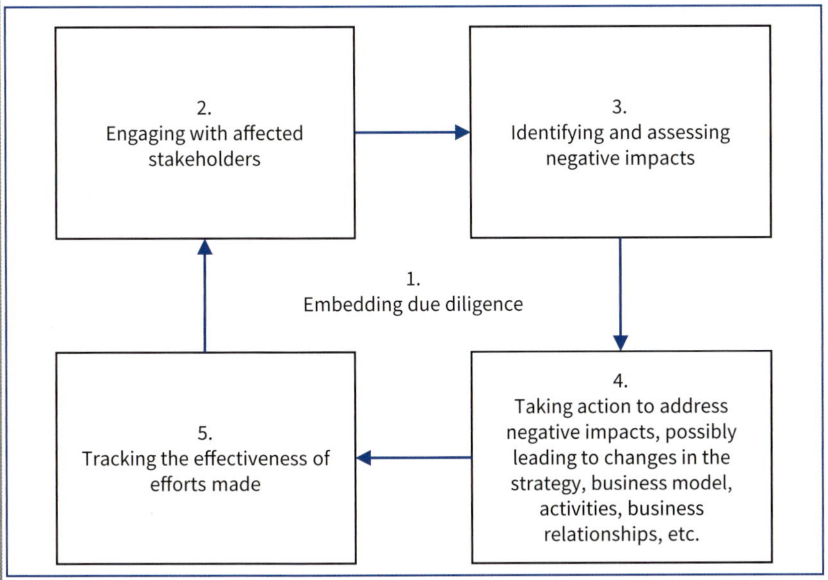

Note:
ESRS 1 refers to the UN Guiding Principles (UNGPs) and the OECD Guidelines for Multinational Enterprises when describing the due diligence process.

The ESRS contain several DRs based on the due diligence process

Process steps of due diligence	**Related DR**
1. Embedding due diligence	• ESRS 2 GOV-2 • ESRS 2 GOV-3 • ESRS 2 SBM-3
2. Engaging with affected stakeholders	• ESRS 2 GOV-2 • ESRS 2 SBM-2 • ESRS 2 IRO-1 • ESRS 2 MDR-P • Topical ESRS (stages and purpose of stakeholder engagement)
3. Identifying and assessing negative impacts	• ESRS 2 IRO-1 • ESRS 2 SBM-3
4. Taking action to address negative impacts	• ESRS 2 MDR-A • Topical ESRS (actions, transition plans)
5. Tracking the effectiveness of efforts made	• ESRS 2 MDR-M • ESRS 2 MDR-T • Topical ESRS (metrics and targets)

[ ] Explanation/Examples

ESRS 1 – General requirements (ESRS 1 – XV/XXIX)

Value chain [5.]

Reporting undertaking and value chain [5.1]

The sustainability statement shall be for the same reporting undertaking as the financial statements, extended to include the value chain

Value chain information	Information on the reporting undertaking shall be extended to include information on the material IROs connected with the undertaking through its direct and indirect business relationships in the upstream and/or downstream value chain. (ESRS 2 VII/XXII)
To be considered when including material IROs connected with the value chain	Material IROs identified through the due diligence process and materiality assessment.
	Specific requirements related to the value chain based on the other ESRS.
Scope of inclusion of value chain information	ESRS do not require information on each and every actor in the value chain, but only the inclusion of material value chain information, which is based on the parts of the value chain for which specific matters are material.
Data of associates or joint ventures that are part of the value chain	Impact metrics are not limited to the share of equity held, but taken into account on the basis of the impacts that are connected with the undertaking's products and services through its business relationships.

XVI/ XXIX

Note:
Disclosures related to policies, actions and targets shall include value chain information where necessary and to the extent that those policies, actions and targets involve actors in the value chain. The focus of most metrics specified by ESRS (without consideration of entity-specific disclosures) is rather on own operations. Even if metrics include value chain information, in some cases, in particular for environmental matters for which proxies are available, the undertaking may be able to comply with the reporting requirements without collecting data directly from the actors in its value chain by using the methods described under [5.2], e.g. when calculating the undertaking's Greenhouse Gas (GHG) Scope 3 emissions.

[- - -] Explanation/Examples

ESRS 1 – General requirements (ESRS 1 – XVI/XXIX)

Value chain [5.]

Estimation using sector averages and proxies [5.2]

Beginning with year four or voluntarily for the first three years of the undertaking's sustainability reporting under the ESRS

After making reasonable efforts, not all the necessary value chain information is available

↓

The undertaking shall estimate the value chain information to be reported by using all reasonable and supportable information without undue cost or effort

Data from indirect sources **or** Sector-average data **or** Sample analyses **or** Market data

or Peer groups data **or** Other proxies **or** Spend-based data **or** Any other reasonable information

↓

Disclosure of policies, actions, metrics and targets including the estimated information

The incorporation of estimates using sector-average data or other proxies shall not result in information that does not meet the qualitative characteristics of information.

VI/XXIX

Note:
Transitional provisions concerning value chain information may be considered.

XXVII/XXIX

Explanation/Examples

ESRS 1 – General requirements (ESRS 1 – XVII/XXIX)

Time horizons [6.]

Reporting period [6.1]

The reporting period for the sustainability statement of the undertaking shall be consistent with that of its financial statements

Linking past, present and future [6.2]

The undertaking shall establish appropriate linkages between retrospective and forward-looking information

Establishment of appropriate linkages	The undertaking shall present how historical information relates to future-oriented information in order to foster a clear understanding.

Reporting progress against the base year [6.3]

Provision of comparative information for targets

Base year	Historical reference date or period for which information is available and against which subsequent information can be compared over time.
Reporting the developments and progress towards a target	The undertaking shall present comparative information in respect of the base year for amounts reported in the current period.

Definition of short-, medium- and long-term for reporting purposes [6.4]

Adaptation of defined time intervals as of the end of the reporting period

Short-term time horizon	Shall be the reporting period.
Medium-term time horizon	From the end of the short-term reporting period to five years.
Long-term time horizon	More than 5 years.
	● The undertaking shall use an **additional breakdown** for the long-term time horizon when impacts or actions are expected in a period longer than five years if necessary to provide relevant information to users of sustainability statements.
Predefined time horizons in topical ESRS	If different definitions of medium- and long-term time horizons are required for specific DRs in other ESRS, the definitions in those ESRS shall prevail.
In case different time horizons are used	Disclosure of the deviating definition of time horizons and reasons for applying those definitions. (ESRS 2 VIII/XXII)

● Requirements that may, if applicable and in combination with other DRs, lead to disclosures dependent on materiality

[ ] Explanation/Examples

ESRS 1 – General requirements (ESRS 1 – XVIII/XXIX)

Preparation and presentation of sustainability information [7.]

Presenting comparative information [7.1]

Disclosure of comparative information in respect of the previous period

Disclosure of comparative information	The undertaking shall disclose comparative information in respect of the previous period for **all quantitative metrics** and **all monetary amounts** disclosed in the current period as well as **narrative disclosures** if relevant to an understanding.
Comparative information that differs from the information reported in the previous year	When reporting comparative information that differs from the information reporting in the previous period, the undertaking shall disclose the **difference** and the **reasons** for the revision of the figure.
If it is impracticable to adjust comparative information for one or more prior periods to achieve comparability	The undertaking shall state this fact.
If more than one comparative period is required by other ESRS	The requirements of the other ESRS shall prevail.

XXVI/XXIX

Note:
Transitional provisions for presenting comparative information [7.1] may be considered.

Sources of estimation and outcome uncertainty [7.2]

When quantitative metrics and monetary amounts cannot be measured directly and can only be estimated, measurement uncertainty may arise

Information to understand uncertainties	ESRS 2 VIII/XXII An undertaking shall disclose information to enable users to understand the most significant uncertainties affecting the quantitative metrics and monetary amounts reported in its sustainability statement.
Consistency of financial data and assumptions	Data and assumptions used in preparing the sustainability statement shall be consistent to the extent possible with the corresponding financial data and assumptions used in the undertaking's financial statements.
Information about uncertain future events	To assess whether such information required by an ESRS is material and should therefore be reported, the undertaking: • Shall refer to the materiality assessment. • Considers the potential financial effects of the events. • Considers the severity and likelihood of the impacts on people or the environment resulting from the possible events. • Considers the full range of possible outcomes and the likelihood of the possible outcomes within that range. • Shall consider all relevant facts and circumstances concerning possible outcomes, including information about low-probability and high-impact outcomes, which, when aggregated, could become material.

● Requirements that may, if applicable and in combination with other DRs, lead to disclosures dependent on materiality

[ ] Explanation/Examples

ESRS 1 – General requirements (ESRS 1 – XIX/XXIX)

Preparation and presentation of sustainability information [7.]

Updating disclosures about events after the end of the reporting period [7.3]

Information received between the end of the reporting period and the approval of the management report for issuance

Information providing evidence or insights about conditions existing at period end	Update estimates and sustainability disclosures in the light of the new information.
Information providing evidence or insights about material transactions, other events and conditions arising after the end of the reporting period ●	Narrative information indicating the existence, nature and potential consequences of these post-year end events, where appropriate.

Changes in preparation or presentation of sustainability information [7.4]

The undertaking shall provide restated comparative figures for the prior period(s) disclosed, if necessary, unless it is impracticable to do so

Consistency	The definition and calculation of metrics shall be consistent over time.
Provision of restated comparative figures, unless it is impracticable to do so	(ESRS 2 IX/XXII) In case of: • Redefined or replaced metrics or targets. • New information in relation to the estimated figures disclosed in the preceding period is identified providing evidence of circumstances that existed in that period.

● Requirements that may, if applicable and in combination with other DRs, lead to disclosures dependent on materiality

[- - -] Explanation/Examples

ESRS 1 – General requirements (ESRS 1 – XX/XXIX)

Preparation and presentation of sustainability information [7.]

Reporting errors in prior periods [7.5]

Prior period errors are omissions from and misstatements in the undertaking's sustainability statement for one or more prior periods (ESRS 2 IX/XXII)

Such errors arise from a failure to use, or misuse of, reliable information that…

- …was available when the management report that includes the sustainability statement for those periods was authorised for issuance.

&

- …could reasonably be expected to have been obtained and considered in the preparation of sustainability disclosures included in these reports.

Examples: The effects of mathematical mistakes, mistakes in applying the definitions for metrics and targets, oversights or misinterpretations of facts and fraud.

If it is impracticable to determine the effect of an error on all prior periods presented	Restatement of the comparative information from the earliest date practicable.
Correction of errors vs. changes in estimates	Corrections of errors are distinguished from changes in estimates. Estimates may need to be revised as additional information becomes known.

Consolidated reporting and subsidiary exemption [7.6]

In case the undertaking is reporting at a consolidated level, the materiality assessment is performed for the entire consolidated group, regardless of its group legal structure

Scope of entities	Covering all subsidiaries in a way that allows for the unbiased identification of material IROs. This includes a top-down or bottom-up approach. (ESRS 2 VII/XXII)
In case of significant differences between material IROs at group level and subsidiary level	● Adequate description of the IROs, as appropriate, of the subsidiary or subsidiaries concerned shall be provided.
When assessing whether those differences are significant	The undertaking may consider different circumstances, such as whether the subsidiary or subsidiaries operate in a different sector than the rest of the group or the circumstances reflected in the level of disaggregation [3.7]. (XII/XXIX)

Note:
The requirement of correcting material prior period errors by restating comparative amounts for the prior period(s) disclosed does not extend to reporting periods before the first year of application of ESRS by the undertaking. [96]

● Requirements that may, if applicable and in combination with other DRs, lead to disclosures dependent on materiality

[- -] Explanation/Examples

ESRS 1 – General requirements (ESRS 1 – XXI/XXIX)

Preparation and presentation of sustainability information [7.]

Classified and sensitive information and information on intellectual property, know-how or results of innovation [7.7]

Omission of a specific piece of information concerning classified or sensitive information or information corresponding to intellectual property, know-how or the results of innovation

Classified information: [Annex II] • As defined in the Council Decision on the security rules for protecting EU classified information or • Classified by one of the Member States and marked as per Appendix B of that Council Decision	The undertaking is not required to disclose such kind of information, even if it is considered material.
Sensitive information as defined in the Regulation establishing the European Defence Fund [Annex II]	The undertaking is not required to disclose such kind of information, even if it is considered material.
Cumulative requirements to omit a specific piece of information corresponding to intellectual property, know-how or the results of innovation	• Applicable for information about strategy, plans and actions, • Information is secret in the sense that it is not generally known among or readily accessible to persons within the circles that normally deal with the kind of information in question, • Information has commercial value because it is secret and • Information has been subject to reasonable steps by the undertaking to keep it secret.
DR in case a piece of information is omitted	Disclose all other required information except for the piece omitted while making every reasonable effort to maintain the relevance of the disclosure.
	The undertaking shall disclose whether it has used the option to omit a specific piece of information. (ESRS 2 VII/XXII)

Explanation/Examples

ESRS 1 – General requirements (ESRS 1 – XXII/XXIX)

Preparation and presentation of sustainability information [7.]

Reporting on opportunities [7.8]

Specific reporting requirements concerning opportunities

Descriptive information	Disclosures concerning opportunities should consist of descriptive information allowing the reader to understand the opportunity for the undertaking or the entire sector.
Consider materiality	The undertaking shall consider: • The materiality of the information to be disclosed. • Whether the opportunity is currently being pursued and is incorporated in its general strategy, as opposed to a general opportunity for the undertaking or the sector. • Whether the inclusion of quantitative measures of anticipated financial effects is appropriate, taking into account the number of assumptions that it could require and consequential uncertainty.

VI/
XXIX

Note:
According to the DRs of ESRS E1 to ESRS E5 related to anticipated financial effects (E1-9, E2-6, E3-5, E4-6, E5-6), a quantification of the financial effects arising from material opportunities is not required if it would result in disclosures that do not meet the qualitative characteristics of information.

[- - -] Explanation/Examples

ESRS 1 – General requirements (ESRS 1 – XXIII/XXIX)

Structure of the sustainability statement [8.]

General presentation requirement [8.1]

General requirements for the presentation of sustainability information

Clear structural distinction	Distinction between information required by disclosures in ESRS and other information included in the management report.
Readability	Structure that facilitates access to and understanding of the sustainability statements, both in human and machine-readable formats.

Note:
ESRS 1 Appendices D and F illustrate examples of the structure of the sustainability statement.

Content and structure of the sustainability statement [8.2]

Details on the structure of the sustainability statement

Dedicated section of the management report	Applicable disclosures required by ESRS shall be reported within a dedicated section of the management report.
Disclosures pursuant to Article 8 of EU Taxonomy	• Shall be included in the sustainability statement in the section "Environmental information". • Shall remain separately identifiable. • Are not subject to the ESRS.
Additional disclosures stemming from other legislation or frameworks (e.g. ISSB, GRI Standards)	VI/ XXIX; ESRS 2 IX/XXII Disclosures shall: • Be clearly identified with an appropriate reference to the related legislation and • Meet the requirements for qualitative characteristics.
Structure of the sustainability statement	The statement shall be structured into four sections in the following order: • General information • Environmental information • Social information • Governance information
Sector-specific ESRS	Such disclosures shall be: • Grouped by reporting area and by sustainability topic. • Presented alongside the disclosures required by ESRS 2 General disclosures and the corresponding topical ESRS.
Entity-specific disclosures	V/ XXIX Such disclosures shall be reported alongside the most relevant sector-agnostic and sector-specific disclosures.

[ ] Explanation/Examples

ESRS 1 – General requirements (ESRS 1 – XXIV/XXIX)

Linkages with other parts of corporate reporting and connected information [9.]

Incorporation by reference [9.1]

Information prescribed by a DR of an ESRS may be incorporated in the sustainability statements by reference

Place of the referenced information	• Another section of the management report • The financial statements • The corporate governance report (if not part of the management report) • The remuneration report • Specific universal registration documents • Specific public disclosures
Cumulative requirements for disclosures incorporated by reference	Incorporated disclosures: • Constitute a separate element of information and are clearly identified in the document concerned as addressing the relevant DRs, or the relevant specific datapoint prescribed by a DR. • Are published before or at the same time as the management report. • Are in the same language as the sustainability statement. • Are subject to at least the same level of assurance as the sustainability statement. • Meet the same technical digitalisation requirements as the sustainability statement. • In case of reference to the EMAS report: use the same basis for preparation, including the scope of consolidation and the treatment of value chain information.
Consideration of overall cohesiveness and readability	When using incorporation by reference, the undertaking shall consider the overall cohesiveness of the reported information and ensure that the incorporation by reference does not impair the readability of the sustainability statements. A list of incorporated information shall be provided. (ESRS 2 IX/XXII)

Note:
ESRS 1 Appendix G illustrates an example of incorporation by reference.

[dashed box] Explanation/Examples

ESRS 1 – General requirements (ESRS 1 – XXV/XXIX)

Linkages with other parts of corporate reporting and connected information [9.]

Connected information and connectivity with financial statements [9.2]

The undertaking shall describe the relationships between different pieces of information

- **Connectivity of information**
  - Connecting narrative information could be required to describe the relationships between governance, strategy and risk management to related metrics and targets.
- **Reference to financial statements in case of direct connectivity**
  - Provide a reference of material monetary amounts or other material quantitative datapoints in the sustainability statement to the corresponding paragraph in the financial statements.
- **Reference to financial statements in case of indirect connectivity**
  - Explain how material monetary amounts or other material quantitative datapoints in the sustainability statement that are an aggregation or part of data presented in the financial statements relate to the most relevant amounts presented in the financial statements, including a reference to the line item and/or relevant paragraphs.
  - Where appropriate, a reconciliation may be provided and it may be presented in a tabular form.
- **Consistency between sustainability statements and financial statements**
  - Explanation of the consistency of significant data, assumptions and qualitative information included in the sustainability statement with the respective information in the financial statements at the level of a single datapoint.
  - When significant inconsistencies arise, the undertaking shall state this to be the fact and explain the reason.
- **Topic-specific connectivity**
  - When a topical/sector-specific ESRS requires reconciliation or an illustration of consistency, this requirement shall prevail.

Requirements that may, if applicable and in combination with other DRs, lead to disclosures dependent on materiality

Requirements that may, if applicable and in combination with other DRs, lead to voluntary disclosures

Explanation/Examples

ESRS 1 – General requirements (ESRS 1 – XXVI/XXIX)

Transitional provisions [10.]

Transitional provisions related to entity-specific disclosures [10.1]

The undertaking may apply transitional provisions for the development of entity-specific disclosures for the first years of reporting (V/XXIX)

Importance of entity-specific disclosures	The extent to which sustainability matters are covered by ESRS is expected to evolve as further DRs are developed, so that entity-specific disclosures will be less necessary in the future. However, they have a high importance as long as sector-specific ESRS have not yet been published in order to ensure faithful reporting.
Transitional measures for entity-specific disclosures	When defining its entity-specific disclosures, the undertaking may adopt transitional measures for their preparation in the first **three** annual sustainability statements.
Entity-specific disclosures reported in prior periods	Report as a priority entity-specific disclosures reported in prior periods, if these disclosures meet or are adapted to meet the qualitative characteristics of information. (VI/XXIX)
Best practice, frameworks and reporting standards	Complement as a priority topical ESRS DRs with sector-specific topics by using available best practice and/or available frameworks or reporting standards, such as industry-based guidance and GRI Sector Standards.

Transitional provisions [10.2] related to the value chain [5.] (XXVII/XXIX)

The undertaking may apply transitional provisions for value chain reporting for the first three years of reporting

Transitional provisions [10.3] related to presenting comparative information [7.1] (XVIII/XXIX)

The undertaking is not required to disclose the comparative information required concerning quantitative metrics, monetary amounts and/or relevant narrative disclosures [7.1] in the **first** year of the sustainability reporting. For phased-in DRs this transitional provision applies with reference to the first year of mandatory application of the phased-in DR

Transitional provision: List of DRs that are phased-in [10.4] (XXVIII/XXIX)

ESRS 1 Appendix C sets phase-in provisions for specific DRs or datapoints of DRs in ESRS that may be omitted or that are not applicable in the first year(s) of preparation of the sustainability statement under the ESRS

[dashed box] Explanation/Examples

ESRS 1 – General requirements (ESRS 1 – XXVII/XXIX)

Transitional provisions [10.]

Transitional provisions related to value chain information [10.2]

XVI/XXIX

For the first three years of the undertaking's sustainability reporting under the ESRS

Not all the necessary value chain information is available

↓

The undertaking shall provide a description of:

- The efforts made to obtain the necessary value chain information
- & The reasons why not all of the necessary information could be obtained
- & The plans to obtain the necessary information in the future

↓

Disclosure of information on policies, actions and targets	Disclosure of information on metrics
The undertaking may limit value chain information to information available in-house, such as data already available to the undertaking and publicly available information.	No value chain information is required, except for datapoints derived from other EU legislation, as listed in [ESRS 2 Appendix B].

Note:
For the first three years, there is no obligation to use estimates if not all the necessary value chain information is available, except for datapoints derived from other EU legislation. However, for policies, actions and targets, ESRS require to use at least in-house data when no value chain information is available within the first three years of sustainability reporting.

● Requirements that may, if applicable and in combination with other DRs, lead to disclosures dependent on materiality

[ ] Explanation/Examples

ESRS 1 – General requirements (ESRS 1 – XXVIII/XXIX)

Transitional provisions [10.]

Appendix C: List of phased-in DRs

ESRS	DR	Paragraph	< 750 employees: omit for the first year	< 750 employees: omit for the first two years	Omit for the first year	Only qualitative disclosures for the first three years
ESRS 2	SBM-1[a]	40 (b), 40 (c)				
	SBM-3	48 (e)			✓	✓[b]
ESRS E1	E1-6	Datapoints on Gross Scopes 1, 2, 3 and Total GHG emissions	✓			
	E1-9	All datapoints			✓	✓[b]
ESRS E2	E2-6	All datapoints			✓	✓[c]
ESRS E3	E3-5	All datapoints			✓	✓
ESRS E4	All DRs	All datapoints		✓ ESRS 2 X/XXII		
	E4-6	All datapoints			✓	✓
ESRS E5	E5-6	All datapoints			✓	✓
ESRS S1	All DRs	All datapoints	✓ ESRS 2 X/XXII			
	S1-7	All datapoints			✓	
	S1-8	All datapoints with regard to own employees in non-EEA countries			✓	
	S1-11	All datapoints			✓	

Note:
(a) The undertaking shall report the information starting from the application date specified in the sector-specific ESRS.
(b) Only qualitative disclosures for the first three years if it is impracticable to prepare quantitative disclosures.
(c) Except for paragraph 40 (b) on the operating and capital expenditures occurred in the reporting period in conjunction with major incidents and deposits.

ESRS 1 – General requirements (ESRS 1 – XXIX/XXIX)

Transitional provisions [10.]

Appendix C: List of phased-in DRs

ESRS	DR	Paragraph	< 750 employees: omit for the first year	< 750 employees: omit for the first two years	Omit for the first year	Only qualitative disclosures for the first three years
ESRS S1	S1-12	All datapoints			✓	
	S1-13	All datapoints			✓	
	S1-14	Datapoints on cases of work-related ill-health and number of days lost to injuries, accidents, fatalities and work-related ill-health			✓	
	S1-14	Reporting on non-employees			✓	
	S1-15	All datapoints			✓	
ESRS S2	All DRs	All datapoints		✓ ESRS 2 X/XXII		
ESRS S3	All DRs	All datapoints		✓ ESRS 2 X/XXII		
ESRS S4	All DRs	All datapoints		✓ ESRS 2 X/XXII		

ESRS 2 (I/I)

Cross-reference between ESRS 2 and topical standards

ESRS Standard	General disclosures						IRO management		Metrics and targets	
	Topical disclosures to be reported alongside ESRS 2 disclosures						MDRs to be applied alongside topical ESRS			
	Governance		Strategy		Processes	Transition plan	Policies	Actions and resources	Metrics	Targets
	GOV-1	GOV-3	SBM-2	SBM-3	IRO-1		MDR-P	MDR-A	MDR-M	MDR-T
E1 Climate change		X		X	X	X	X	X	X	X
E2 Pollution					X		X	X	X	X
E3 Water and marine resources					X		X	X	X	X
E4 Biodiversity and ecosystems				X	X	X	X	X	X	X
E5 Resource use and circular economy					X		X	X	X	X
S1 Own workforce			X	X			X	X	X	X
S2 Workers in the value chain			X	X			X	X	X	X
S3 Affected communities			X	X			X	X	X	X
S4 Consumers and end-users			X	X			X	X	X	X
G1 Business conduct	X				X		X	X	X	X

Intentionally left blank

ESRS 2 – General disclosures

Objective:

Enabling understanding of:

- The DRs that apply to all undertakings regardless of their sector of activity and across all sustainability matters.

Sustainability matters: All

Interaction with other ESRS:

- **Conjunctions**: The general disclosures of ESRS 2 shall be applied to all topical and sector-specific ESRS as well as to entity-specific disclosures.
- **Intersectional topics covered by other standards**: n/a

Interaction with other frameworks:

- Accounting Directive

Overall evaluation:

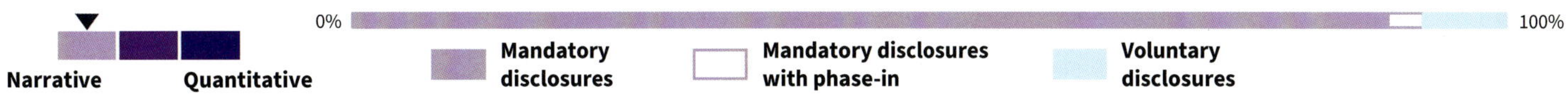

The overall evaluation contains a subjective assessment of the corresponding standard as narrative or quantitative and a count of all disclosures included in the corresponding standard. The number of disclosures was determined using professional judgement and depends on various factors. Due to the subjectivity involved, other counts may deviate.

ESRS 2 – General disclosures (ESRS 2 – I/XXII)

Structure and content

Basis for preparation

BP-1 (VII/XXII)

General basis for preparation of the sustainability statement

Information on how the undertaking prepares its sustainability statement.

BP-2 (VIII/XXII)

Disclosures in relation to **specific circumstances**

Information on the effects of specific circumstances on the preparation of the sustainability statement.

Strategy

SBM-1

Strategy, business model and value chain

The key elements of the general strategy that relate to or affect sustainability matters, business model and value chain.

SBM-2 (XII/XXII)

Interests and views of stakeholders

Information on how the interests and views of stakeholders inform the undertaking's strategy and business model.

SBM-3 (XIII/XXII)

Material IROs and their interaction with **strategy and business model**

Information on material IROs and how they interact with strategy and business model.

Governance

GOV-1 (XI/XXII)

The **role of the administrative, management and supervisory bodies (AMSB)**

Composition and diversity of the AMSB and their roles, responsibilities and (access to) expertise and skills.

GOV-2 (XI/XXII)

Information provided to and **sustainability matters addressed** by the undertaking's **AMSB**

Information on how the AMSB are informed about sustainability matters and how these were addressed during the reporting period.

GOV-3 (XI/XXII)

Integration of sustainability-related performance in **incentive schemes**

Whether and how incentive schemes offered to the AMSB are linked to sustainability matters.

GOV-4 (XI/XXII)

Statement on due diligence

Information on the due diligence process in form of a mapping that explains how and where the application of the main aspects and steps of the due diligence process are reflected in the sustainability statement.

GOV-5 (XI/XXII)

Risk management and internal controls over sustainability reporting

Information on the main features of risk management and internal control processes in relation to the sustainability reporting process.

Impact, risk and opportunity management

IRO-1 (XV/XXII)

Description of the **process** to **identify** and **assess** material **IROs**

Information on the process to identify and assess material IROs based on the principle of double materiality.

IRO-2

DRs in ESRS **covered** by the sustainability statement

Information on the DRs included in the sustainability statement and of the topics that have been omitted as not material.

Policies MDR-P (XVII/XXII)

Policies adopted to manage material sustainability matters

Summarised information on the key content, scope and communication of the policies in place to manage IROs.

Actions MDR-A (XVIII/XXII)

Actions and **resources** in relation to material sustainability matters

Summarised information on key actions taken and/or planned and resources to manage IROs.

Metrics and targets

Metrics MDR-M (XX/XXII)

Metrics in relation to material sustainability matters

Summarised information on metrics in place with regard to each material sustainability matter.

Targets MDR-T (XXI/XXII)

Tracking effectiveness of policies and actions through **targets**

Summarised information on targets and progress regarding IROs.

ESRS 2 – General disclosures (ESRS 2 – II/XXII)

Key disclosures concerning the basis for preparation

VIII/XXII

Disclosures in relation to specific circumstances [BP-2]

- Identification of metrics that include value chain data estimated by using indirect sources, e.g. sector-average data or other proxies [10 (a)]
- Description of the resulting level of accuracy of metrics that include value chain data estimated by using indirect sources [10 (c)]
- Identification of quantitative metrics and monetary amounts disclosed that are subject to a high level of measurement uncertainty [11 (a)]
- In relation to each quantitative metric and monetary amount identified under [11 (a)]: information about the sources of measurement uncertainty [11 (b) (i)]
- Extent to which data and processes used for sustainability reporting purposes have been verified by an external assurance provider and found to conform to the corresponding standard (e.g. ISO) [AR 2]

Mandatory disclosure Voluntary disclosure

ESRS 2 – General disclosures (ESRS 2 – III/XXII)

Key disclosures concerning governance – 1/2

The role of the AMSB [GOV-1] (XI/XXII)

- Number of executive and number of non-executive members of AMSB [21 (a)]
- Representation of employees and other workers in AMSB [21 (b)]
- Experience of the AMSB relevant to the sectors, products and geographic locations of the undertaking [21 (c)]
- Percentage by gender of AMSB [21 (d)]
- Percentage by other diversity aspects considered of AMSB [21 (d)]
- Share of independent board members of AMSB [21 (e)]

Breakdowns: Administrative bodies | Management bodies | Supervisory bodies

Information provided to and sustainability matters addressed by the undertaking's AMSB [GOV-2] (XI/XXII)

- Whether, by whom and how frequently the AMSB are informed about: [26 (a)]
  - Material IROs
  - The implementation of due diligence
  - The results and effectiveness of policies, actions, metrics and targets
- List of the material IROs addressed by the AMSB or their relevant committees during the reporting period [26 (c)]

Mandatory disclosure | Breakdowns

ESRS 2 – General disclosures (ESRS 2 – IV/XXII)

Key disclosures concerning governance – 2/2

Integration of sustainability-related performance in incentive schemes [GOV-3] (XI/XXII)

- Whether performance is being assessed against specific sustainability-related targets and/or impacts and, if so, which ones [29 (b)]
- Proportion of variable remuneration of the AMSB dependent on sustainability-related targets and/or impacts [29 (d)]

Risk management and internal controls over sustainability reporting [GOV-5] (XI/XXII)

- Main risks identified and their mitigation strategies in the risk management and internal control system in relation to the sustainability reporting process, including related controls [36 (c)]

Mandatory disclosure

ESRS 2 – General disclosures (ESRS 2 – V/XXII)

Key disclosures concerning strategy

Strategy, business model and value chain [SBM-1]

- Headcount of employees [40 (a) (iii)]
  - Geographical areas
- Products and services that are banned in certain markets [40 (a) (iv)]
- Revenue [40 (b)]
  - Significant ESRS sectors
- Statement indicating, together with the related revenues, that the undertaking is active in: [40 (d)]
  - Fossil fuel (coal, oil and gas) sector and revenues derived from EU Taxonomy-aligned economic activities related to fossil gas
  - Chemicals production
  - Controversial weapons
  - Cultivation and production of tobacco
- Description of the undertaking's business model and value chain, including: [42]
  - Inputs and approach to gathering, developing and securing those inputs
  - Outputs and outcomes in terms of current and expected benefits for customers, investors and other stakeholders
  - For key value chains: main features of the value chain, including the undertaking's position and main business actors and their relationship to the undertaking

Material IROs and their interaction with strategy and business model [SBM-3]

XIII/XXII

- Brief description of the material IROs resulting from the materiality assessment, including where the IROs are concentrated [48 (a)]
- Current and anticipated effects of the material IROs on the business model, value chain, strategy and decision-making and how the undertaking has responded or plans to respond to these effects [48 (b)]
- How the material impacts affect or are likely to affect people or the environment [48 (c) (i)]
- Current financial effects of the material risks and opportunities on the financial position, financial performance and cash flows [48 (d)]
- Anticipated financial effects of the material risks and opportunities on the financial position, financial performance and cash flows, including expected changes, given its strategy to manage risks and opportunities [48 (e)]
  - Short-term
  - Medium-term
  - Long-term
- Qualitative and, when applicable, quantitative analysis of the resilience of the strategy and business model regarding its capacity to manage material IROs [48 (f)]

Mandatory disclosure | Mandatory disclosure with phase-in | Breakdowns

ESRS 2 – General disclosures (ESRS 2 – VI/XXII)

Key disclosures concerning IRO management

XV/ XXII

Description of the process to identify and assess material IROs [IRO-1]

Extent to which and how the process to identify, assess and manage impacts and risks is integrated into the overall risk management process and used to evaluate the risk profile and risk management processes [53 (e)]

Extent to which and how the process to identify, assess and manage opportunities is integrated into the overall management process [53 (f)]

Mandatory disclosure

ESRS 2 – General disclosures (ESRS 2 – VII/XXII)

Disclosure Requirement ESRS 2 BP-1 | General basis for preparation of the sustainability statement

Provide an understanding of how the undertaking prepares its sustainability statement. [4]

Consolidated or individual reporting

- Whether the sustainability statement has been prepared on a consolidated or individual basis [5 (a)]

Consolidated statements (ESRS 1 XX/XXIX)

- Confirmation that the scope of consolidation is the same as for the financial statements [5 (b) (i)]
- If applicable, declaration that the reporting undertaking is not required to draw up financial statements [5 (b) (i)]
- If applicable, declaration that the reporting undertaking is preparing consolidated sustainability reporting pursuant to Article 48i of the Accounting Directive [5 (b) (i)]
- Indication of which subsidiaries included in the consolidation are exempted from sustainability reporting [5 (b) (ii)]

Coverage of value chain (ESRS 1 XV/XXIX)

- Extent to which the sustainability statement covers the value chain [5 (c)]
  - Concerning the materiality assessment [AR 1 (a)]
  - Concerning the policies, actions and targets [AR 1 (b)]
  - Concerning the metrics [AR 1 (c)]

Options for omitting information

- Whether the undertaking has used the option to omit a specific piece of information corresponding to intellectual property, know-how or the results of innovation [5 (d)] (ESRS 1 XXI/XXIX)
- Whether the undertaking has made use of the exemption from disclosing impending developments or matters in the course of negotiation [5 (e)]
  - Member state option resulting from Article 19a paragraph 3 or Article 29a paragraph 3 of the Accounting Directive, whose applicability depends on the transposition into national law

Mandatory disclosure | Explanation/Examples

ESRS 2 – General disclosures (ESRS 2 – VIII/XXII)

Disclosure Requirement ESRS 2 BP-2 | Disclosures in relation to specific circumstances – 1/3

Provide an understanding of the effect of specific circumstances on the preparation of the sustainability statement. [7]

Time horizons (ESRS 1 XVII/XXIX)

- Definitions of time horizons deviating from ESRS 1 for medium- or long-term time horizons [9 (a)]
  - Reasons for application of deviating definitions of time horizons [9 (b)]

Value chain estimation (ESRS 1 XVI/XXIX)

- Identification of metrics that include value chain data estimated using indirect sources (e.g. sector-average data, other proxies) [10 (a)]
  - Basis for preparation of those metrics [10 (b)]
- Resulting level of accuracy of those metrics [10 (c)]
  - Planned actions to improve the accuracy of those metrics in the future [10 (d)]

Sources of estimation and outcome uncertainty (ESRS 1 XVIII/XXIX)

- Identification of quantitative metrics and monetary amounts disclosed with a high level of measurement uncertainty [11 (a)]
  - Information about the sources of measurement uncertainty concerning identified metrics and monetary amounts [11 (b) (i)]
- Assumptions, approximations and judgements in measuring the identified metrics and monetary amounts [11 (b) (ii)]
  - Indication that forward-looking information is considered to be uncertain [12]

Note:
These disclosures may be reported alongside the disclosures to which they refer. [8]

Mandatory disclosure | Voluntary disclosure | Explanation/Examples

ESRS 2 – General disclosures (ESRS 2 – IX/XXII)

Disclosure Requirement ESRS 2 BP-2 | Disclosures in relation to specific circumstances – 2/3

Changes in preparation or presentation of sustainability information (ESRS 1 XIX/ XXIX)

- Explanation of changes, reasons for those and why the replaced metrics provide more useful information [13 (a)]
- Revised comparative figures or statement that it is impracticable to adjust this prior period information [13 (b)]
- Difference between the figure disclosed in the preceding period and the revised comparative figure [13 (c)]

Reporting errors in prior periods (ESRS 1 XX/ XXIX)

- Nature of the prior period material error [14 (a)]
- Correction for each prior period to the extent practicable [14 (b)]
- If not practicable, explanation of the circumstances that led to this condition [14 (c)]

Disclosures stemming from other legislation or other pronouncements (ESRS 1 XXIII/ XXIX)

- Statement that information stemming from other legislation or generally accepted sustainability reporting standards and frameworks is disclosed [15]
- In case of partial application: precise reference to the paragraphs of the standard or framework applied [15]
- Whether the undertaking relies on any European standards approved by the European Standardisation System (e.g. ISO) [AR 2] *(voluntary)*
- Extent to which data and processes that are used for sustainability reporting purposes have been verified by an external assurance provider and found to conform to the corresponding standard (e.g. ISO) [AR 2] *(voluntary)*

Incorporation by reference (ESRS 1 XXIV/ XXIX)

- List of ESRS DRs or datapoints that have been incorporated by reference [16]

Note:
These disclosures may be reported alongside the disclosures to which they refer. [8]

Mandatory disclosure | Voluntary disclosure

ESRS 2 – General disclosures (ESRS 2 – X/XXII)

Disclosure Requirement ESRS 2 BP-2 | Disclosures in relation to specific circumstances – 3/3

Use of phase-in provisions for small undertakings

An undertaking or group not exceeding on its balance sheet date the average number of 750 employees during the financial year (on a consolidated basis, where applicable) may omit the information required by ESRS E4, ESRS S1, ESRS S2, ESRS S3 or ESRS S4 for the first or first two years of preparation of its sustainability statement.
[ESRS 1 Appendix C]

≤ 750 may omit

First year	First two years	
ESRS S1	ESRS E4	ESRS S2
	ESRS S3	ESRS S4

In this case the undertaking shall nevertheless disclose...

- Whether the relevant sustainability topics have been assessed to be material [17]
- List of material sustainability matters according to the list of matters in [ESRS 1.AR 16] [17 (a)]
- Brief description of how the business model and strategy take account of the impacts related to each material sustainability matter [17 (a)]
- Brief description of any time-bound targets set in relation to each material sustainability matter, the progress made and whether the targets related to biodiversity and ecosystems are based on conclusive scientific evidence [17 (b)]
- Brief description of policies in relation to each material sustainability matter [17 (c)]
- Brief description of actions taken in relation to each material sustainability matter to identify, monitor, prevent, mitigate, remediate or bring an end to actual or potential adverse impacts and the result of such actions [17 (d)]
- Metrics relevant to each material sustainability matter [17 (e)]

Note:
Concerning the disclosures presented under ESRS 2.17 no phase-ins are applicable, as these requirements directly result from the Accounting Directive.

Disclosure dependent on materiality

Explanation/Examples

ESRS 2 – General disclosures (ESRS 2 – XI/XXII)

Disclosure Requirements ESRS 2 GOV-1, GOV-2, GOV-3, GOV-4 and GOV-5 | Governance processes

The DRs ESRS 2 GOV-1 to GOV-5 aim to enable an understanding of the governance processes, controls and procedures put in place to monitor, manage and oversee sustainability matters, [18] e.g.:

Description of management's role in the governance processes, controls and procedures used to monitor, manage and oversee IROs [GOV-1, 22 (c)]

Sustainability-related expertise that the AMSB, as a whole, either directly possess or can leverage to oversee sustainability matters, e.g. through access to experts or training [GOV-1, 23 (a)]

Risk assessment approach followed in the risk management and internal control system in relation to the sustainability reporting process, including risk prioritisation [GOV-5, 36 (b)]

Main risks identified in the risk management and internal control system in relation to the sustainability reporting process, including related controls [GOV-5, 36 (c)]

How the AMSB and senior executive management oversee the setting of targets related to material IROs and how they monitor progress towards them [GOV-1, 22 (d)]

Control environment

Risk assessment

Control activities

Information and communication

Monitoring activities

Whether, by whom and how frequently the AMSB are informed about material IROs, the implementation of due diligence and the results and effectiveness of policies, actions, metrics and targets adopted to address the IROs [GOV-2, 26 (a)]

Description of the periodic reporting of the findings of the risk assessment and internal controls regarding the sustainability reporting process to the AMSB [GOV-5, 36 (e)]

How the AMSB consider IROs when overseeing the undertaking's strategy, its decisions on major transactions and its risk management process [GOV-2, 26 (b)]

How the governance bodies ensure that an appropriate mechanism for performance monitoring is in place [GOV-2, AR 6]

Mandatory disclosure | Voluntary disclosure | Explanation/Examples

ESRS 2 – General disclosures (ESRS 2 – XII/XXII)

Disclosure Requirement ESRS 2 SBM-2 | Interests and views of stakeholders

Engagement with stakeholders

Summarised description of the undertaking's stakeholder engagement, including: [45 (a)]

- Key stakeholders
- Whether engagement with them occurs and for which categories of stakeholders
- How the engagement is organised
- The purpose of the engagement
- How the outcome is taken into account by the undertaking

Understanding of interests and views of stakeholders

Summarised description of the interests and views of key stakeholders as they relate to the undertaking's strategy and business model [45 (b)]

- To the extent that these were analysed during the due diligence process and/or materiality assessment process

Amendment of strategy and business model

Where applicable, summarised description of amendments to the strategy and/or business model, including: [45 (c)]

- How the undertaking has amended or expects to amend its strategy and/or business model to address the interests and views of its stakeholders
- Any further steps that are being planned and in what timeline
- Whether these steps are likely to modify the relationship with and views of stakeholders

Information of the AMSB

Whether and how the AMSB are informed about the views and interests of affected stakeholders regarding the undertaking's impacts [45 (d)]

Mandatory disclosure Explanation/Examples

ESRS 2 – General disclosures (ESRS 2 – XIII/XXII)

Disclosure Requirement ESRS 2 SBM-3 | Material IROs and their interaction with strategy and business model – 1/2

Disclosures concerning material IROs

- Brief description of material IROs [48 (a)]
- Where in the undertaking's business model, own operations and value chain the material IROs are concentrated [48 (a)]
- Changes to material IROs compared to the previous reporting period [48 (g)]
- IROs covered by ESRS and IROs covered by entity-specific disclosures [48 (h)]

Disclosures concerning strategy and business model

- Current and anticipated effects of the material IROs on business model, value chain, strategy and decision-making [48 (b)]
- How the undertaking has responded or plans to respond to the effects, including any changes it has made or plans to make to its strategy or business model to manage material IROs [48 (b)]
- Information about the resilience of the strategy and business model regarding its capacity to manage material IROs [48 (f)]
- Qualitative and, when applicable, quantitative analysis of the resilience of the strategy and business model, including contextual information [48 (f)]

Note:
The disclosures may be reported alongside the disclosures provided under the corresponding topical ESRS, in which case the undertaking shall still present a statement of its material IROs alongside its disclosures prepared under the chapter "General information".
This disclosure may be expressed in terms of a single IRO or by aggregating groups of material IROs when this provides more relevant information and does not obscure material information. [AR 18]

Mandatory disclosure

ESRS 2 – General disclosures (ESRS 2 – XIV/XXII)

Disclosure Requirement ESRS 2 SBM-3 | Material IROs and their interaction with strategy and business model – 2/2

Disclosures concerning material impacts | **Disclosures concerning material risks and opportunities**

Reasonably expected time horizons
[48 (c) (iii), (e)]

Disclosures concerning material impacts

- How the material impacts affect or are likely to affect people or the environment [48 (c) (i)]
- Whether and how the material impacts originate from or are connected to the undertaking's strategy and business model [48 (c) (ii)]
- Whether the undertaking is involved through its activities or because of its business relationships and description of the nature of those [48 (c) (iv)]

Disclosures concerning material risks and opportunities

- Current financial effects of the material risks and opportunities on the financial position, financial performance and cash flows [48 (d)]
- Material risks and opportunities for which there is a significant risk of a material adjustment within the next annual reporting period to the carrying amounts of assets and liabilities [48 (d)]
- Anticipated financial effects of the material risks and opportunities on the financial position, financial performance and cash flows, including expected changes, given its strategy to manage risks and opportunities [48 (e)]
  - Short-term
  - Medium-term
  - Long-term

Mandatory disclosure | Mandatory disclosure with phase-in | Breakdowns

ESRS 2 – General disclosures (ESRS 2 – XV/XXII)

Disclosure Requirement ESRS 2 IRO-1 | Description of the process to identify and assess material IROs – 1/2

Description of the methodologies and assumptions applied in the materiality process [53 (a)], input parameters used [53 (g)], whether and how the process has changed compared to the prior period, when the process was modified for the last time and future revision dates of the materiality assessment [53 (h)]

Process to identify and assess impacts [53 (b)]

Overview of the process to identify, assess, prioritise and monitor the potential and actual impacts on people and the environment, informed by the undertaking's due diligence process

- Whether and how the process focusses on specific activities, business relationships, geographies or other factors that give rise to heightened risk of adverse impacts
- Whether and how the process considers the impacts with which the undertaking is involved through its own operations or as a result of its business relationships
- Whether and how the process includes consultation with affected stakeholders and with external experts
- Whether and how the process prioritises negative and positive impacts based on their materiality factors
- Whether and how the process determines which matters are material for reporting purposes (including thresholds and other criteria used)

Process to identify and assess risks and opportunities [53 (c)]

Overview of the process used to identify, assess, prioritise and monitor risks and opportunities that have or may have financial effects

- How the undertaking has considered the connections of its impacts and dependencies with the risks and opportunities
- How the undertaking assesses the materiality factors and nature of effects of the identified risk and opportunities (e.g. thresholds and other criteria used)
- How the undertaking prioritises sustainability-related risks relative to other types of risks, including its use of risk assessment tools

Integration into internal processes

Description of the decision-making process and related internal controls [53 (d)]

Extent to which and how the process to identify, assess and manage impacts and risks is integrated into the overall risk management process and used to evaluate the overall risk profile and risk management process [53 (e)]

Extent to which and how the process to identify, assess and manage opportunities is integrated into the overall management process [53 (f)]

Note:
In addition to the mandatory disclosures of ESRS 2 IRO-1, the topical specifications in ESRS E1 to ESRS E5 and ESRS G1 related to DR IRO-1 shall be reported in all instances, independently of the materiality of the sustainability matter. [2 (a)]

Mandatory disclosure

ESRS 2 – General disclosures (ESRS 2 – XVI/XXII)

Disclosure Requirement ESRS 2 IRO-1 | Description of the processes to identify and assess material IROs – 2/2

Description of the process to identify and assess material IROs [51], including topical information as specified in the E- and G-Standards, e.g.:

Whether the undertaking has screened e.g. its assets and activities to identify
- Climate-related IROs [ESRS E1.20 (a), (b), (c)]
- Pollution-related IROs [ESRS E2.11 (a)],
- Water and marine resources-related IROs [ESRS E3.8 (a)]
- Resource use and circular economy-related IROs [ESRS E5.11 (a)]

Whether and how the undertaking has identified and assessed impacts on biodiversity and ecosystems at its own site locations and in the value chain, transition and physical risks and opportunities [ESRS E4.17 (a), (c)]

All relevant criteria used in the process to identify material IROs, including location, activity, sector and structure of the transaction [ESRS G1.6]

The materiality assessment process concerning ESRS E2, E3, E4 and E5 **may** follow the **LEAP approach**: [ESRS E2.AR 1, E3.AR 1, E4.AR 6, E5.AR 1]

L

Phase 1: **Locate** where in own operations and along the value chain the interface with nature takes place

Consider e.g.:
- (Site) Locations of direct assets and operations and activities across the value chain
- Sites located in sensitive/impacted areas
- Sectors or business units related to the material matters

E

Phase 2: **Evaluate** the dependencies and impacts

Assess the materiality of the impacts for each material site, sector/business or priority location by using the materiality factors:
- Severity
- Likelihood

A

Phase 3: **Assess** the material risks and opportunities

Identify e.g.:
- Transition risks (policy and legal, technology, market, reputation)
- Physical risks (acute and chronic)
- Systemic risks
- Opportunities (business and sustainability performance)

P

Phase 4: **Prepare** and report the results of the materiality assessment

Consider e.g.:
- A list of site locations, business units or areas associated with the material IROs
- A list of business activities, commodities or resources associated with the material IROs

Mandatory disclosure | Explanation/Examples

ESRS 2 – General disclosures (ESRS 2 – XVII/XXII)

Minimum Disclosure Requirement MDR-P | Policies adopted to manage material sustainability matters

Disclosures concerning the policies the undertaking has in place with regard to each material sustainability matter

- Description of the key contents of the policy, including its general objectives, the related material IROs and the process for monitoring [65 (a)]
- Description of the scope of the policy, or its exclusions, in terms of activities, value chain, geographies and, if relevant, affected stakeholder groups [65 (b)]
- Explanation of: [AR 21]
  - The activities and/or segments of own operations or value chain the policy concerns
  - Further boundaries relevant to the specific topic or the circumstances
  - Extent of the value chain covered by the policy
- Most senior level in the undertaking's organisation that is accountable for implementing the policy [65 (c)]
- Reference to third-party standards or initiatives the undertaking commits to respect through the implementation of the policy [65 (d)]
- If the undertaking cannot disclose the information on policies required under relevant ESRS, because it has not adopted policies with reference to the specific sustainability matter concerned, it shall disclose this to be the case and provide reasons for not having adopted policies [62]
  - Timeframe in which the undertaking aims to adopt such policies concerning the material sustainability matter [62]
- Description of consideration given to the interests of key stakeholders in setting the policy [65 (e)]
- Whether and how the policy is made available to potentially affected stakeholders and stakeholders who need to help implement it [65 (f)]

Note:
The disclosures shall be located alongside the DRs in the relevant topical ESRS and, if applicable, alongside entity-specific disclosures. [60, 61] The disclosures based on ESRS 2 MDR-P may be supplemented by topic-specific disclosures, which are defined in the topical ESRS.
In some cases, the DR of the ESRS cannot be assigned as a whole and clearly to one of the areas of policies, actions, metrics or targets.
In these cases, the DR were assigned to the area to which they predominantly correspond. An individual assessment at the level of the data-points to identify the relevant MDR may be necessary.
When a single policy addresses several interconnected sustainability matters, the undertaking may disclose the required information in its reporting under one topical ESRS and cross reference to it in its reporting under the other topical ESRS. [61]

Disclosure dependent on materiality | Voluntary disclosure

Intentionally left blank

ESRS 2 – General disclosures (ESRS 2 – XVIII/XXII)

Minimum Disclosure Requirement MDR-A | Actions and resources in relation to material sustainability matters – 1/2

Disclosures concerning the actions through which the undertaking manages each material sustainability matter

Information on key actions

Not every implemented action needs to be reported when the standard refers to "key actions". Key actions are those actions that materially contribute to achieving the undertaking's objectives in addressing material IROs. For reasons of understandability, key actions may be aggregated where appropriate. [AR 22]

List of key actions taken and planned, their expected outcomes and their contribution to the achievement of policy objectives and targets [68 (a)]

Time horizons under which each key action is intended to be completed [68 (c)]

If the undertaking cannot disclose the information on actions required under relevant ESRS, because it has not adopted actions with reference to the specific sustainability matter concerned, it shall disclose this to be the case and provide reasons for not having adopted actions [62]

Timeframe in which the undertaking aims to adopt such actions concerning the material sustainability matter [62]

Scope of the key actions [68 (b)]

Key actions taken (along with results) to provide for and cooperate in or support the provision of remedy for those harmed by actual material impacts [68 (d)]

Quantitative and qualitative information regarding the progress of actions or action plans disclosed in prior periods [68 (e)]

Note:
The disclosures shall be located alongside the DRs in the relevant topical ESRS and, if applicable, alongside entity-specific disclosures. [60, 61] The disclosures based on ESRS 2 MDR-A may be supplemented by topic-specific disclosures, which are defined in the topical ESRS.
In some cases, the DR of the ESRS cannot be assigned as a whole and clearly to one of the areas of policies, actions, metrics or targets. In these cases, the DR were assigned to the area to which they predominantly correspond. An individual assessment at the level of the datapoints to identify the relevant MDR may be necessary.

Disclosure dependent on materiality | Voluntary disclosure | Explanation/Examples

ESRS 2 – General disclosures (ESRS 2 – XIX/XXII)

Minimum Disclosure Requirement MDR-A | Actions and resources in relation to material sustainability matters – 2/2

Where the implementation of an action plan requires significant OpEx and/or CapEx

Description of the type of current and future financial and other resources allocated to the action plan, including the relevant terms of sustainable finance instruments and the environmental or social objectives
[69 (a)]

Whether the ability to implement the action or action plan depends on specific preconditions
[69 (a)]

E.g. granting of financial support or public policy and market developments

Amount of current and amount of future financial resources for the implementation of a key action or action plan
[69 (b), (c)]

How the amount of current financial resources relates to the most relevant amounts presented in the financial statements
[69 (b)]

CapEx
[AR 23]

OpEx
[AR 23]

Across the relevant time horizons
[AR 23]

Resources applied in the reporting year
[AR 23]

Planned allocation of resources over specific time horizons
[AR 23]

Note:
When the same action addresses several interconnected sustainability matters, the undertaking may disclose the required information in its reporting under one topical ESRS and cross reference to it in its reporting under the other topical ESRS. [61]

Disclosure dependent on materiality | Voluntary disclosure | Breakdowns | Explanation/Examples

ESRS 2 – General disclosures (ESRS 2 – XX/XXII)

Minimum Disclosure Requirement MDR-M | Metrics in relation to material sustainability matters

Disclosures concerning the metrics the undertaking has in place with regard to each material sustainability matter

- Any metrics used to evaluate performance and effectiveness in relation to a material IRO [75]
  - Metrics defined in ESRS [76]
  - Metrics identified on an entity-specific basis, whether taken from other sources or developed by the undertaking itself [76] (ESRS 1 V/XXIX)
  - Methodologies and significant assumptions behind the metric, including the limitations of the methodologies used [77 (a)]
  - Whether the measurement of the metric is validated by an external body other than the assurance provider and, if so, which body [77 (b)]
  - Labelling and definition of the metric using meaningful, clear and precise names and descriptions [77 (c)]
    - When currency is specified as the unit of measure, the presentation currency of the financial statements shall be used [77 (d)]

Note:
The disclosures shall be located alongside the DRs in the relevant topical ESRS and, if applicable, alongside entity-specific disclosures. [70, 71] The disclosures based on ESRS 2 MDR-M may be supplemented by topic-specific disclosures, which are defined in the topical ESRS.
In some cases, the DR of the ESRS cannot be assigned as a whole and clearly to one of the areas of policies, actions, metrics or targets. In these cases, the DR were assigned to the area to which they predominantly correspond. An individual assessment at the level of the datapoints to identify the relevant MDR may be necessary.

Disclosure dependent on materiality | Explanation/Examples

Intentionally left blank

ESRS 2 – General disclosures (ESRS 2 – XXI/XXII)

Minimum Disclosure Requirement MDR-T | Tracking effectiveness of policies and actions through targets – 1/2

Disclosures concerning the targets set with regard to each material sustainability matter

Measurable, outcome-oriented and time-bound targets set [80]

- Relationship of the target to the policy objectives [80 (a)]
- Defined target level to be achieved [80 (b)]
- Whether the target is absolute or relative and unit of measurement [80 (b)]
  - Targets related to the prevention or mitigation of environmental impacts in absolute terms shall be prioritised over corresponding targets in relative terms. [AR 24]
  - Targets addressing the prevention or mitigation of social impacts may be specified in terms of the effects on human rights, welfare or positive outcomes for affected stakeholders. [AR 24]
- Scope, including activities, value chain and geographical boundaries [80 (c)]
- Baseline value and base year from which progress is measured [80 (d)]
- Period to which the target applies and any milestones or interim targets [80 (e)]
- Methodologies and significant assumptions used, including scenarios, data sources, alignment with policy goals [80 (f)]
- How the target considers the wider context of sustainable development and/or local situation in which impacts take place [80 (f)]
- Whether environment-related targets are based on conclusive scientific evidence [80 (g)]
- Whether and how stakeholders have been involved in the target setting [80 (h)]

Note:
The disclosures shall be located alongside the DRs in the relevant topical ESRS and, if applicable, alongside entity-specific disclosures. [70, 71] The disclosures based on ESRS 2 MDR-T may be supplemented by topic-specific disclosures, which are defined in the topical ESRS.
In some cases, the DR of the ESRS cannot be assigned as a whole and clearly to one of the areas of policies, actions, metrics or targets. In these cases, the DR were assigned to the area to which they predominantly correspond. An individual assessment at the level of the datapoints to identify the relevant MDR may be necessary.

Disclosure dependent on materiality (shaded boxes) | Explanation/Examples (dashed boxes)

ESRS 2 – General disclosures (ESRS 2 – XXII/XXII)

Minimum Disclosure Requirement MDR-T | Tracking effectiveness of policies and actions through targets – 2/2

Assess progress and changes over time

Any changes in: [80 (i)]
- Targets and corresponding metrics
- Measurement methodologies
- Significant assumptions
- Limitation, sources and processes to collect data adopted within the defined time horizon
- Including an explanation of the rational for changes and their effects on comparability

Performance against the disclosed targets, including an analysis of trends or significant changes in the performance
[80 (j)]

The information on progress made towards achieving the targets may be presented in a comprehensive table, including information on the baseline and target value, milestones and achieved performance over the prior periods.
[AR 25]

Information on how the target is monitored and reviewed and the metrics used and whether progress is in line with what had been initially planned
[80 (j)]

If no measurable, outcome-oriented targets have been set

If the undertaking cannot disclose the information on targets required under relevant ESRS, because it has not set targets with reference to the specific sustainability matter concerned, it shall disclose this to be the case and provide reasons for not having adopted targets
[72]

Timeframe in which the undertaking aims to adopt such targets concerning the material sustainability matter
[72]

Whether measurable outcome-oriented targets will be set and the timeframe for setting them or reasons for not setting them
[81 (a)]

Whether the effectiveness of policies and actions is nevertheless being tracked
[81 (b)]

If nevertheless the effectiveness of policies and actions is being tracked

Any processes to track effectiveness of policies and actions
[81 (b) (i)]

Any qualitative or quantitative indicators used to evaluate progress, including the base period
[81 (b) (ii)]

Defined level of ambition to be achieved
[81 (b) (ii)]

Baseline against which the progress is considered
[AR 26]

Disclosure dependent on materiality | Voluntary disclosure | Explanation/Examples

5 Environment

ESRS E1 – E5 (I/III)

Overview of sustainability matters of ESRS E-Standards – 1/2

Overview of ESRS VIII/XIII

	Climate change E1 (E1-I/XVI)		Pollution E2 (E2-I/VII)		Water and marine resources E3 (E3-I/X)	
Topics	**Climate change E1**		**Pollution E2**		**Water and marine resources E3**	
Sub-topics	Climate change mitigation and adaptation	Energy	Pollution of: • Air • Water • Soil • Living organisms and food resources	Substances of concern and very high concern and microplastics	Water	Marine resources
Sub-sub-topics					• Water consumption • Water withdrawals • Water discharges • Water discharges in the oceans • Extraction and use of marine resources	

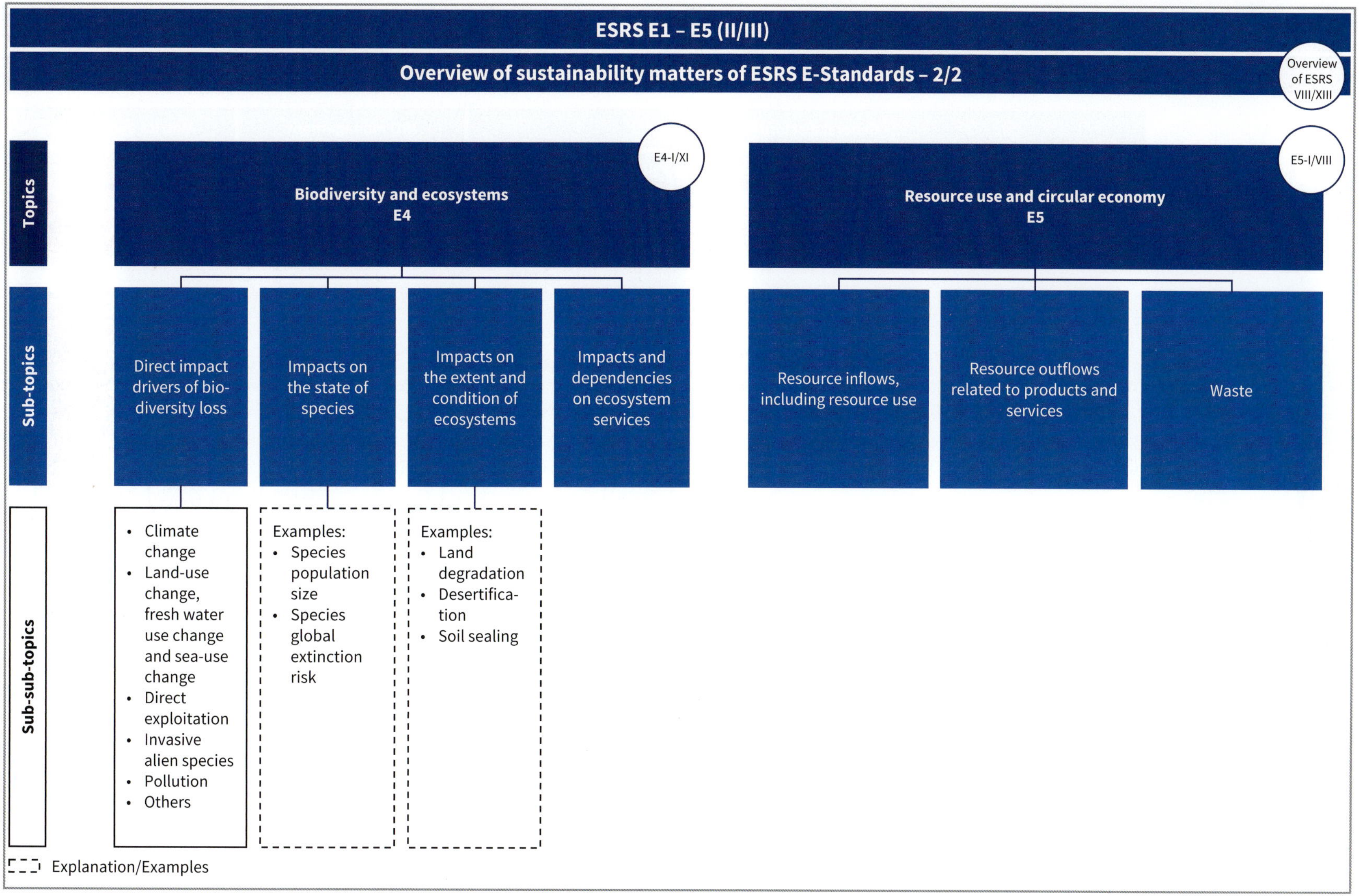
ESRS E1 – E5 (II/III)
Overview of sustainability matters of ESRS E-Standards – 2/2
Overview of ESRS VIII/XIII
Topics
Sub-topics
Sub-sub-topics
E4-I/XI
Biodiversity and ecosystems E4
Direct impact drivers of bio-diversity loss
Impacts on the state of species
Impacts on the extent and condition of ecosystems
Impacts and dependencies on ecosystem services
• Climate change
• Land-use change, fresh water use change and sea-use change
• Direct exploitation
• Invasive alien species
• Pollution
• Others
Examples:
• Species population size
• Species global extinction risk
Examples:
• Land degradation
• Desertifica-tion
• Soil sealing
E5-I/VIII
Resource use and circular economy E5
Resource inflows, including resource use
Resource outflows related to products and services
Waste
Explanation/Examples

ESRS E1 – E5 (III/III)

Cross-reference between ESRS 2 and ESRS E-Standards

		General disclosures						IRO management		Metrics and targets	
		Topical disclosures to be reported alongside ESRS 2 disclosures						MDRs to be applied alongside topical ESRS			
		Governance		Strategy		Processes	Transition plan	Policies	Actions and resources	Metrics	Targets
		GOV-1	GOV-3	SBM-2	SBM-3	IRO-1		MDR-P	MDR-A	MDR-M	MDR-T
ESRS Standard	**E1** Climate change		E1-II/XVI		E1-II/XVI	E1-III/XVI	E1-II/XVI	E1-I/XVI	E1-IV/XVI	E1-VI/XVI E1-XIII/XVI	E1-V/XVI
	E2 Pollution					E2-I/VII		E2-II/VII	E2-II/VII	E2-III/VII E2-V/VII	E2-III/VII
	E3 Water and marine resources					E3-I/X		E3-I/X E3-VI/X	E3-II/X E3-VII/X	E3-IV/X E3-IX/X	E3-III/X E3-VIII/X
	E4 Biodiversity and ecosystems				E4-II/XI	E4-II/XI	E4-II/XI E4-VII/XI	E4-III/XI	E4-III/XI	E4-V/XI E4-IX/XI	E4-IV/XI
	E5 Resource use and circular economy					E5-I/VIII		E5-I/VIII	E5-II/VIII	E5-III/VIII E5-VI/VIII	E5-III/VIII E5-V/VIII

ESRS E1 – Climate change

Objective:	Enabling understanding of: • The material positive and negative actual and potential impacts of the undertaking on climate change. • The undertaking's past, current and future mitigation efforts in line with the Paris Agreement or an updated international agreement on climate change and compatible with limiting global warming to 1.5°C. • The plans and capacity of the undertaking to adapt its strategy and business model, in line with the transition to a sustainable economy and to contribute to limiting global warming to 1.5°C. • Any other actions taken and the result of such actions to prevent, mitigate or remediate actual or potential negative impacts and to address risks and opportunities. • The nature, type and extent of the undertaking's material risks and opportunities arising from its impacts and dependencies on climate change and how it manages them. • The financial effects on the undertaking over the short-, medium- and long-term of material risks and opportunities arising from its impacts and dependencies on climate change.
Sustainability matters:	Climate change mitigation, climate change adaptation, energy
Interaction with other ESRS:	
• **Conjunctions**	ESRS 1 (General requirements), ESRS 2 (General disclosures)
• **Intersectional topics covered by other standards**	ESRS E2 (Ozone-depleting substances, nitrogen oxides and sulphur oxides, among other air emissions that are related to climate change) ESRS E4 (Biodiversity loss and ecosystem degradation caused by climate change) ESRS S1, ESRS S2, ESRS S3, ESRS S4 (Impacts on people from the transition to a climate-neutral economy)
Interaction with other frameworks:	• Delegated Regulation supplementing the EU Taxonomy by establishing technical screening criteria • Delegated Regulation supplementing the EU Taxonomy by specifying the content and presentation of information • EU Climate Law • EU Taxonomy • GHG Protocol Corporate Standard • GHG Protocol Corporate Value Chain Standard • GRI 305 • ISO 14064 Quantification and reporting of greenhouse gas emissions and removals
Applicability:	Undertakings or groups not exceeding the average number of 750 employees may omit the datapoints on scope 3 emissions and total GHG emissions specified in the DR ESRS E1-6 for the first year of preparation of their sustainability statement.

Overall evaluation:

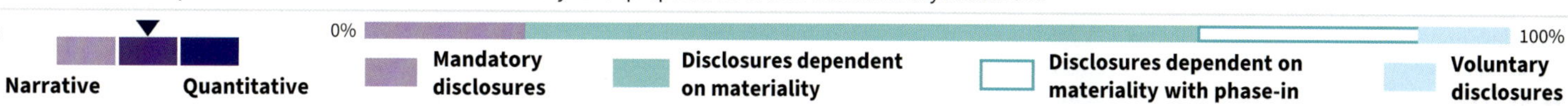

The overall evaluation contains a subjective assessment of the corresponding standard as narrative or quantitative and a count of all disclosures included in the corresponding standard. The number of disclosures was determined using professional judgement and depends on various factors. Due to the subjectivity involved, other counts may deviate.

ESRS E1 – Climate change (E1 – I/XVI)

Structure and content

General disclosures

ESRS 2 GOV-3

Integration of sustainability-related performance in **incentive schemes**

Information on whether and how climate-related considerations are factored into the remuneration of members of the AMSB against GHG emission reduction targets and the share of the remuneration linked to climate related considerations, including their explanation.

E1-1

Transition plan for climate change mitigation

Information on whether and how strategy and business model are compatible with the transition to a sustainable economy and with the limiting of global warming to 1.5 °C and achieving climate neutrality by 2050 in line with the Paris Agreement.

ESRS 2 SBM-3

Material IROs and their interaction with **strategy and business model**

Categorisation of risks into climate-related physical risks or climate-related transition risks and information on the resilience analysis including the use of climate scenario analysis and its results.

ESRS 2 IRO-1

Description of the **processes** to **identify** and **assess** material climate-related **IROs**

Information on the processes to identify and assess climate-related impacts, physical risks and transitional risks and opportunities.

Impact, risk and opportunity management

E1-2 (ESRS 2 MDR-P)

Policies related to climate change mitigation and adaptation

Information on the extent to which policies address the identification, assessment, management and/or remediation of material climate change mitigation and adaptation IROs.

E1-3 (ESRS 2 MDR-A)

Actions and **resources** related to climate change policies

Information on key actions taken and planned and the resources allocated to their implementation to achieve climate-related policy objectives and targets.

Metrics and targets

E1-4 (ESRS 2 MDR-T)

Targets related to climate change mitigation and adaptation

Information on targets set to support climate change mitigation and adaptation policies and address material climate-related IROs.

E1-5 (ESRS 2 MDR-M)

Energy consumption and mix

Information on the total energy consumption in absolute values, improvement in energy efficiency, exposure to coal, oil and gas-related activities and the share of renewable energy in the energy mix.

E1-6 (ESRS 2 MDR-M) (XIII/XVI)

Gross scopes 1, 2, 3 and Total GHG emissions

Information on the undertaking's GHG emissions (scope 1, scope 2, scope 3 and total emissions) as well as the GHG intensity (GHG emissions per net revenue).

E1-7 (ESRS 2 MDR-M) (XV/XVI)

GHG removals and mitigation projects financed through **carbon credits**

Information on GHG removals and storage resulting from projects developed in own operations or contributed to in the value chain, as well as information on the amount of GHG reduction or removals through carbon credits.

E1-8 (ESRS 2 MDR-M) (XVI/XVI)

Internal carbon pricing

Information on the internal carbon pricing scheme including the applied carbon prices, how they support policies and targets and the GHG emission volumes covered by those schemes.

E1-9 (ESRS 2 MDR-M)

Anticipated financial effects

Information on anticipated financial effects due to material physical and transition risks and potential climate-related opportunities, including assets and net revenue at material physical and transition risk as well as cost savings related to climate change mitigation.

ESRS E1 – Climate change (E1 – II/XVI)

Key disclosures concerning general disclosures – 1/2

Integration of sustainability-related performance in incentive schemes [ESRS 2 GOV-3]

- Share of remuneration recognised in the current period that is linked to climate-related considerations [13]

Transition plan for climate change mitigation [E1-1]

- Quantification of investments and funding supporting the implementation of the transition plan [16 (c)]
- Qualitative assessment of the potential locked-in GHG emissions from key assets and products [16 (d)]
- Significant CapEx amounts invested during the reporting period related to: [16 (f)]
  - Coal-related economic activities
  - Oil-related economic activities
  - Gas-related economic activities
- Explanation of the progress in implementing the transition plan [16 (j)]

Material IROs and their interaction with strategy and business model [ESRS 2 SBM-3]

- Whether each material climate-related risk is considered to be a physical or transition risk [18]
- Description of the result of the resilience analysis of the strategy and business model in relation to climate change, including the results from the use of climate scenario analysis [19 (c)]
- Which part of own operations and the value chain as well as which material physical risks and transition risks may have been excluded from the resilience analysis [AR 6]

Disclosure dependent on materiality (all boxes above)

ESRS E1 – Climate change (E1 – III/XVI)

Key disclosures concerning general disclosures– 2/2

Description of the processes to identify and assess material climate-related IROs [ESRS 2 IRO-1]

Whether and how:

- The extent to which assets and business activities may be exposed and are sensitive to the identified climate-related hazards has been assessed [AR 11 (c)]
- Transition events have been identified over the short-, medium- and long-term [AR 12 (a)]
- The extent to which assets and business activities may be exposed and are sensitive to the identified transition events has been assessed [AR 12 (b)]
- Assets and business activities that are incompatible with or need significant efforts to be compatible with a transition to a climate-neutral economy have been identified [AR 12 (d)]

Which scenarios were used in the climate-related scenario analysis, their sources and alignment with state-of-the-art science [AR 13 (a)]

Mandatory disclosure

ESRS E1 – Climate change (E1 – IV/XVI)

Key disclosures concerning IRO management

ESRS 2 MDR-A

Actions and resources related to climate change policies [E1-3]

- List of key actions taken and planned, their expected outcomes and their contribution to the achievement of policy objectives and targets [ESRS 2 MDR-A, 68 (a)]
- Key actions taken (along with results) to provide remedy for those harmed by actual material impacts [ESRS 2 MDR-A, 68 (d)]
- Whether the ability to implement each key action depends on specific preconditions [ESRS 2 MDR-A, 69 (a)]
- Amount of current and future financial resources for the implementation of a key action or action plan [ESRS 2 MDR-A, 69 (b), (c)]
- Achieved and expected GHG emission reductions from key actions [29 (b)]

Disclosure dependent on materiality

ESRS E1 – Climate change (E1 – V/XVI)

Key disclosures concerning metrics and targets – 1/8

ESRS 2 MDR-T

Targets related to climate change mitigation and adaptation [E1-4]

- Performance against each disclosed target, including an analysis of trends or significant changes in the performance [ESRS 2 MDR-T, 80 (j)]
- Information on how each target is monitored and reviewed and whether progress is in line with what had been initially planned [ESRS 2 MDR-T, 80 (j)]
- Overall estimated quantitative contributions of expected decarbonisation levers to achieve GHG emission reduction targets [34 (f)]
  - Scope 1 [AR 30 (a)]
  - Scope 2 [AR 30 (a)]
  - Scope 3 [AR 30 (a)]
- Whether GHG emission reduction targets are science-based and compatible with limiting global warming to 1.5°C [34 (e)]

Disclosure dependent on materiality

Breakdowns

ESRS E1 – Climate change (E1 – VI/XVI)

Key disclosures concerning metrics and targets – 2/8

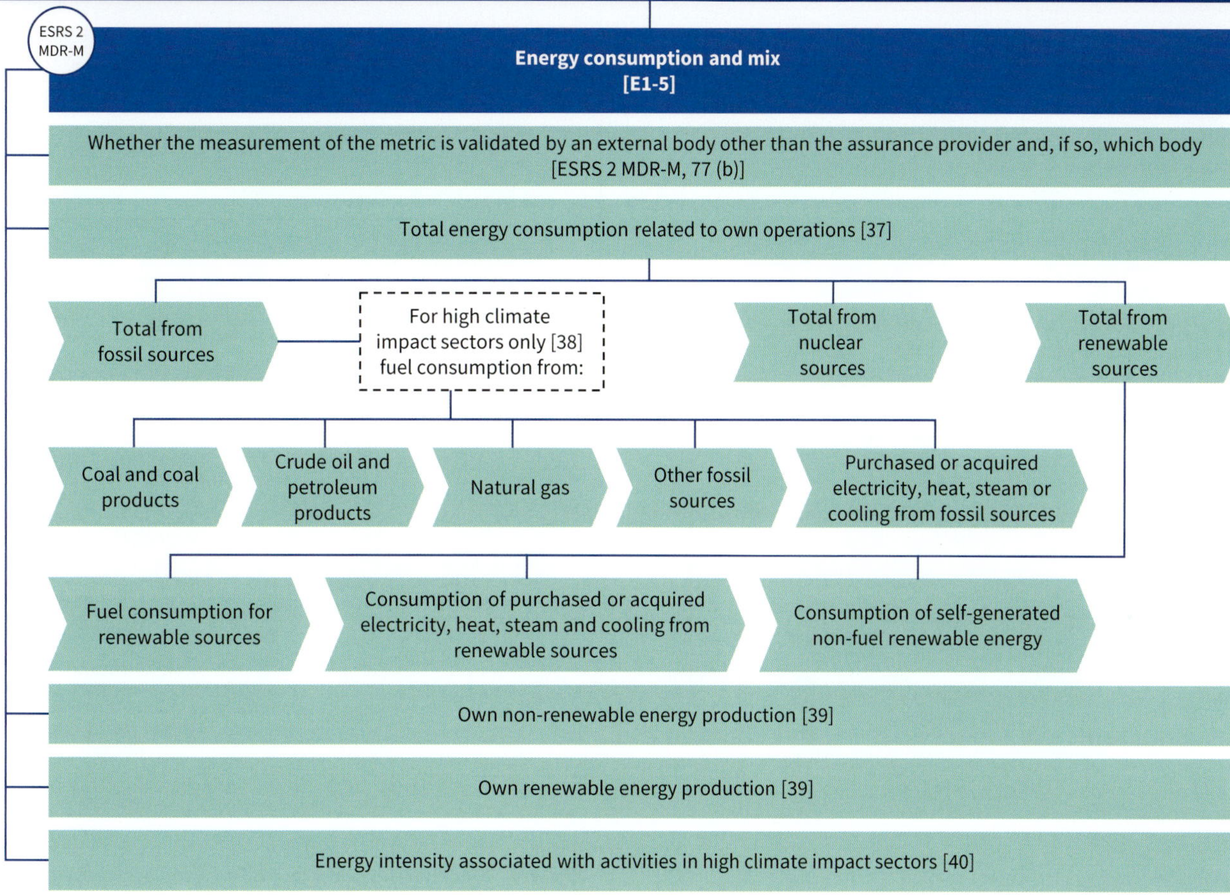

Disclosure dependent on materiality | Breakdowns | Explanation/Examples

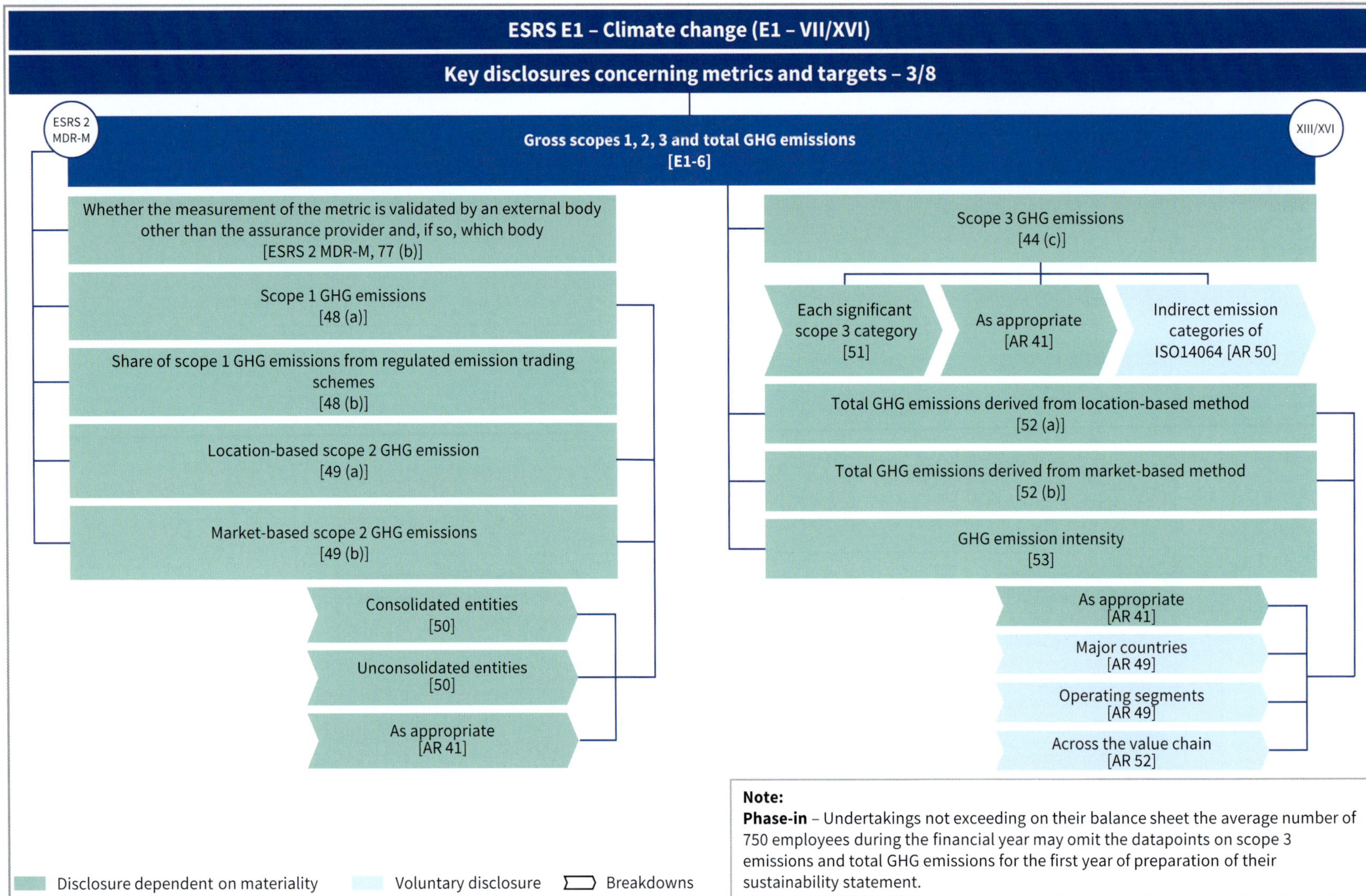
ESRS E1 – Climate change (E1 – VII/XVI)
Key disclosures concerning metrics and targets – 3/8
ESRS 2 MDR-M
XIII/XVI
Gross scopes 1, 2, 3 and total GHG emissions [E1-6]
Whether the measurement of the metric is validated by an external body other than the assurance provider and, if so, which body [ESRS 2 MDR-M, 77 (b)]
Scope 1 GHG emissions [48 (a)]
Share of scope 1 GHG emissions from regulated emission trading schemes [48 (b)]
Location-based scope 2 GHG emission [49 (a)]
Market-based scope 2 GHG emissions [49 (b)]
Consolidated entities [50]
Unconsolidated entities [50]
As appropriate [AR 41]
Scope 3 GHG emissions [44 (c)]
Each significant scope 3 category [51]
As appropriate [AR 41]
Indirect emission categories of ISO14064 [AR 50]
Total GHG emissions derived from location-based method [52 (a)]
Total GHG emissions derived from market-based method [52 (b)]
GHG emission intensity [53]
As appropriate [AR 41]
Major countries [AR 49]
Operating segments [AR 49]
Across the value chain [AR 52]
Note:
Phase-in – Undertakings not exceeding on their balance sheet the average number of 750 employees during the financial year may omit the datapoints on scope 3 emissions and total GHG emissions for the first year of preparation of their sustainability statement.
Disclosure dependent on materiality
Voluntary disclosure
Breakdowns

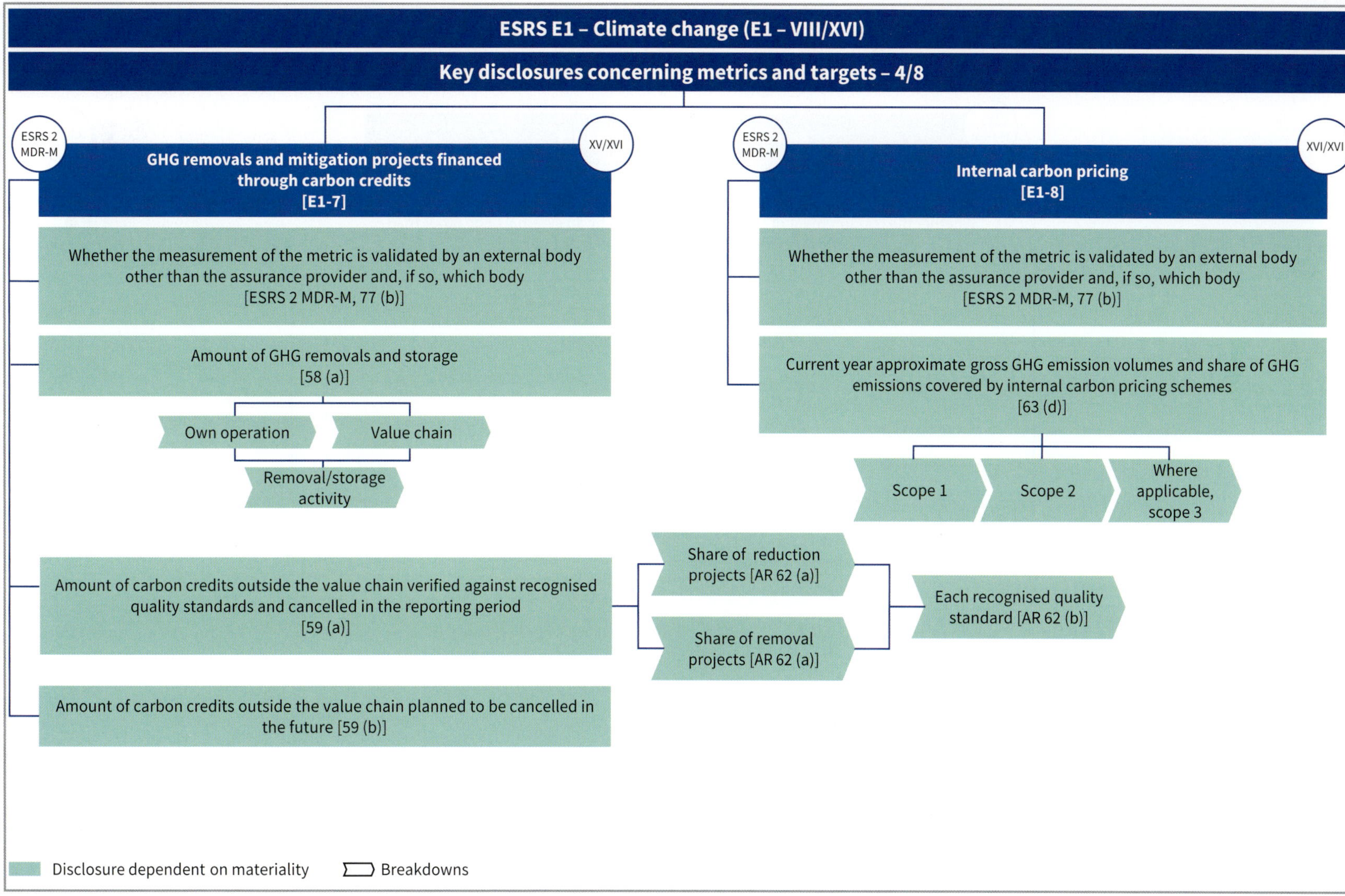
ESRS E1 – Climate change (E1 – VIII/XVI)
Key disclosures concerning metrics and targets – 4/8
ESRS 2 MDR-M
GHG removals and mitigation projects financed through carbon credits [E1-7]
XV/XVI
Whether the measurement of the metric is validated by an external body other than the assurance provider and, if so, which body [ESRS 2 MDR-M, 77 (b)]
Amount of GHG removals and storage [58 (a)]
Own operation
Value chain
Removal/storage activity
Amount of carbon credits outside the value chain verified against recognised quality standards and cancelled in the reporting period [59 (a)]
Share of reduction projects [AR 62 (a)]
Share of removal projects [AR 62 (a)]
Each recognised quality standard [AR 62 (b)]
Amount of carbon credits outside the value chain planned to be cancelled in the future [59 (b)]
ESRS 2 MDR-M
Internal carbon pricing [E1-8]
XVI/XVI
Whether the measurement of the metric is validated by an external body other than the assurance provider and, if so, which body [ESRS 2 MDR-M, 77 (b)]
Current year approximate gross GHG emission volumes and share of GHG emissions covered by internal carbon pricing schemes [63 (d)]
Scope 1
Scope 2
Where applicable, scope 3
Disclosure dependent on materiality
Breakdowns

ESRS E1 – Climate change (E1 – IX/XVI)

Key disclosures concerning metrics and targets – 5/8

ESRS 2 MDR-M

Anticipated financial effects – 1/4
[E1-9]

- Whether the measurement of the metric is validated by an external body other than the assurance provider and, if so, which body [ESRS 2 MDR-M, 77 (b)]
- Monetary amount of assets at material physical risk before considering climate change adaptation actions [66 (a)]
  - Acute risk
  - Chronic risk
- Share of assets at material physical risk before considering climate change adaptation actions [66 (a)]
- Share of assets at material physical risk addressed by climate change adaptation actions [66 (b)]

Breakdowns (for monetary amount and share of assets at material physical risk before considering climate change adaptation actions):

- Short-term
- Medium-term
- Long-term

Disclosure dependent on materiality with phase-in — Breakdowns

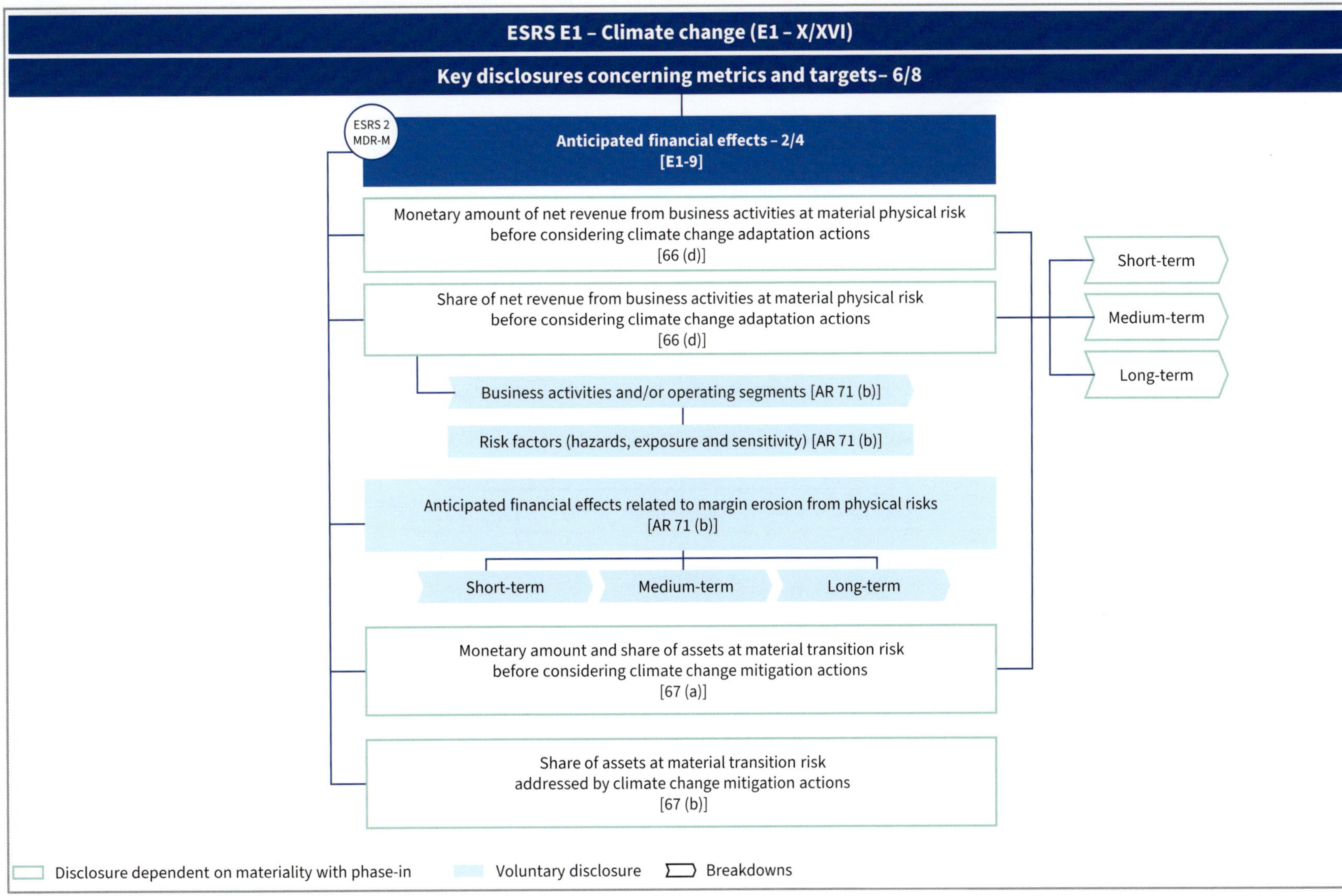
ESRS E1 – Climate change (E1 – X/XVI)
Key disclosures concerning metrics and targets– 6/8
ESRS 2 MDR-M
Anticipated financial effects – 2/4 [E1-9]
Monetary amount of net revenue from business activities at material physical risk before considering climate change adaptation actions [66 (d)]
Share of net revenue from business activities at material physical risk before considering climate change adaptation actions [66 (d)]
Short-term
Medium-term
Long-term
Business activities and/or operating segments [AR 71 (b)]
Risk factors (hazards, exposure and sensitivity) [AR 71 (b)]
Anticipated financial effects related to margin erosion from physical risks [AR 71 (b)]
Short-term
Medium-term
Long-term
Monetary amount and share of assets at material transition risk before considering climate change mitigation actions [67 (a)]
Share of assets at material transition risk addressed by climate change mitigation actions [67 (b)]
Disclosure dependent on materiality with phase-in
Voluntary disclosure
Breakdowns

ESRS E1 – Climate change (E1 – XI/XVI)

Key disclosures concerning metrics and targets – 7/8

Anticipated financial effects – 3/4 [E1-9] (ESRS 2 MDR-M)

- Carrying value of real-estate assets [67 (c)]
  - Energy-efficiency classes
- Estimated amount and share of potentially stranded assets [AR 73 (a)]
  - From the reporting year until 2030
  - From 2030 to 2050
- Liabilities that may have to be recognised in financial statements due to material transition risks [67 (d)]
  - Short-term
  - Medium-term
  - Long-term
- Range of potential future liabilities originating from emission trading schemes [AR 74 (a)]
- Potential future liabilities that relate to allocation plans for the period before and until 2030 [AR 74 (b)]
- Number of scope 1 GHG emission allowances within regulated emission schemes [AR 74 (c)]
- Cumulative number of scope 1 GHG emission allowances stored at the beginning of the reporting period [AR 74 (c)]
- Potential future liabilities associated with carbon credits planned to be cancelled in the near future based on existing contractual agreements [AR 74 (d)]

Disclosure dependent on materiality with phase-in | Voluntary disclosure | Breakdowns

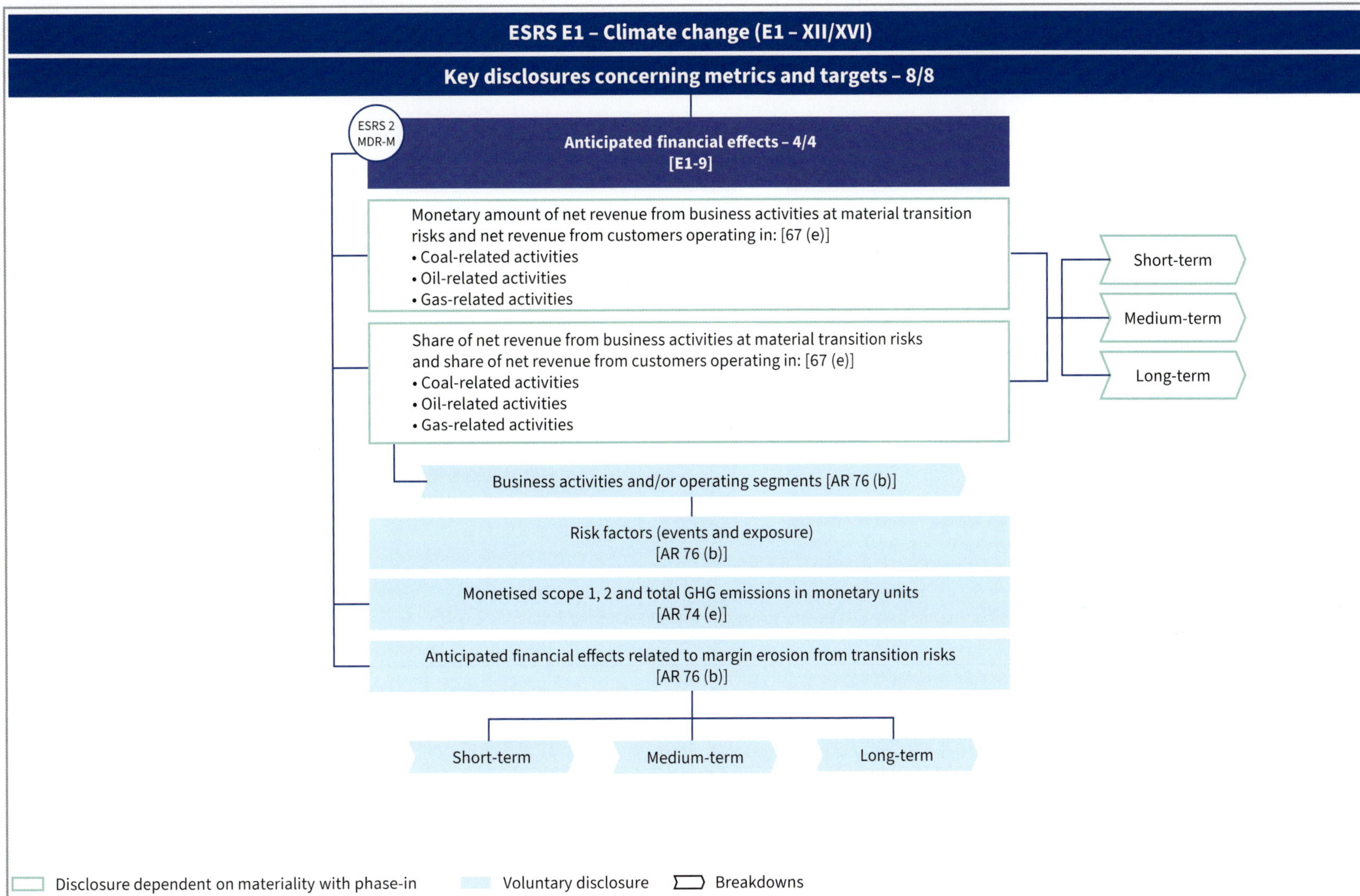
ESRS E1 – Climate change (E1 – XII/XVI)
Key disclosures concerning metrics and targets – 8/8
ESRS 2 MDR-M
Anticipated financial effects – 4/4 [E1-9]
Monetary amount of net revenue from business activities at material transition risks and net revenue from customers operating in: [67 (e)]
• Coal-related activities
• Oil-related activities
• Gas-related activities
Share of net revenue from business activities at material transition risks and share of net revenue from customers operating in: [67 (e)]
• Coal-related activities
• Oil-related activities
• Gas-related activities
Short-term
Medium-term
Long-term
Business activities and/or operating segments [AR 76 (b)]
Risk factors (events and exposure) [AR 76 (b)]
Monetised scope 1, 2 and total GHG emissions in monetary units [AR 74 (e)]
Anticipated financial effects related to margin erosion from transition risks [AR 76 (b)]
Short-term
Medium-term
Long-term
Disclosure dependent on materiality with phase-in
Voluntary disclosure
Breakdowns

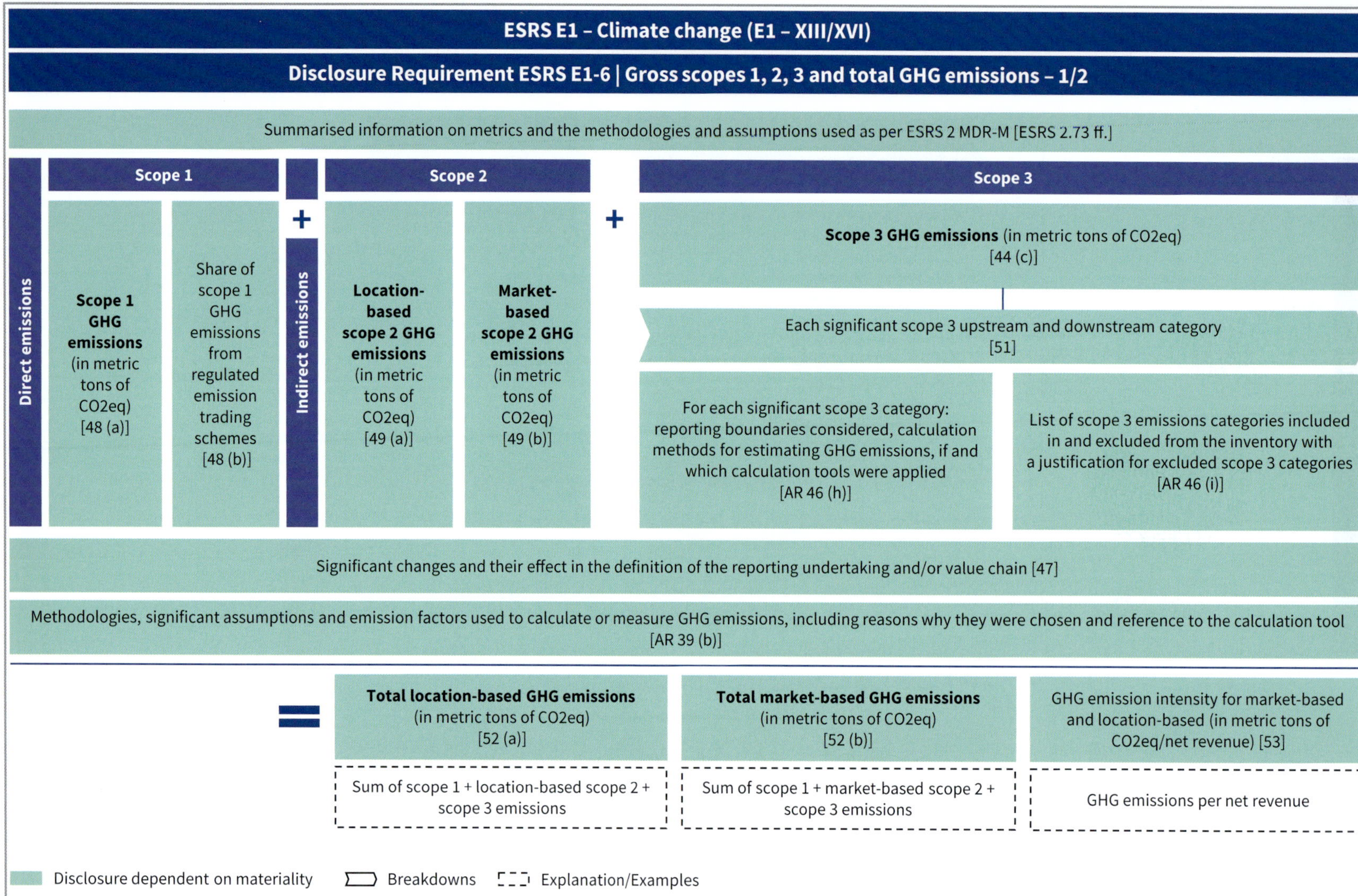
ESRS E1 – Climate change (E1 – XIII/XVI)
Disclosure Requirement ESRS E1-6 | Gross scopes 1, 2, 3 and total GHG emissions – 1/2
Summarised information on metrics and the methodologies and assumptions used as per ESRS 2 MDR-M [ESRS 2.73 ff.]
Direct emissions
Scope 1
Scope 1 GHG emissions (in metric tons of CO2eq) [48 (a)]
Share of scope 1 GHG emissions from regulated emission trading schemes [48 (b)]
+
Indirect emissions
Scope 2
Location-based scope 2 GHG emissions (in metric tons of CO2eq) [49 (a)]
Market-based scope 2 GHG emissions (in metric tons of CO2eq) [49 (b)]
+
Scope 3
Scope 3 GHG emissions (in metric tons of CO2eq) [44 (c)]
Each significant scope 3 upstream and downstream category [51]
For each significant scope 3 category: reporting boundaries considered, calculation methods for estimating GHG emissions, if and which calculation tools were applied [AR 46 (h)]
List of scope 3 emissions categories included in and excluded from the inventory with a justification for excluded scope 3 categories [AR 46 (i)]
Significant changes and their effect in the definition of the reporting undertaking and/or value chain [47]
Methodologies, significant assumptions and emission factors used to calculate or measure GHG emissions, including reasons why they were chosen and reference to the calculation tool [AR 39 (b)]
=
Total location-based GHG emissions (in metric tons of CO2eq) [52 (a)]
Sum of scope 1 + location-based scope 2 + scope 3 emissions
Total market-based GHG emissions (in metric tons of CO2eq) [52 (b)]
Sum of scope 1 + market-based scope 2 + scope 3 emissions
GHG emission intensity for market-based and location-based (in metric tons of CO2eq/net revenue) [53]
GHG emissions per net revenue
Disclosure dependent on materiality
Breakdowns
Explanation/Examples

ESRS E1 – Climate change (E1 – XIV/XVI)

Disclosure Requirement ESRS E1-6 | Gross scopes 1, 2, 3 and total GHG emissions – 2/2

Explanations

Direct emissions

Scope 1

Undertaking activities

- Emissions from the undertaking facilities (e.g. heating, steam, cooling)
- Emissions from the undertaking vehicles

+

Indirect emissions

Scope 2

Undertaking activities

Emissions from the generation of purchased or acquired heat, steam, cooling and electricity consumed by the undertaking

Location-based = Emissions based on the average energy generation emission factors for defined locations	Market-based = Emissions emitted by the generators from which the undertaking contractually purchases electricity

+

Scope 3

Upstream categories

- Business travels
- Capital goods
- Operational waste
- Employee commuting
- Fuel- and energy-related activities
- Upstream transportation and distribution
- Purchased goods and services (including cloud computing and data centre services)
- Upstream leased assets

Downstream categories

- Investments
- Franchises
- Use phase
- Downstream transportation and distribution
- Processing of sold products
- End-of-life treatment of sold products
- Downstream leased assets

= **Total emissions**

Note:
Phase-in – Undertakings not exceeding on their balance sheet the average number of 750 employees during the financial year may omit the datapoints on scope 3 emissions and total GHG emissions for the first year of preparation of their sustainability statement.

Explanation/Examples

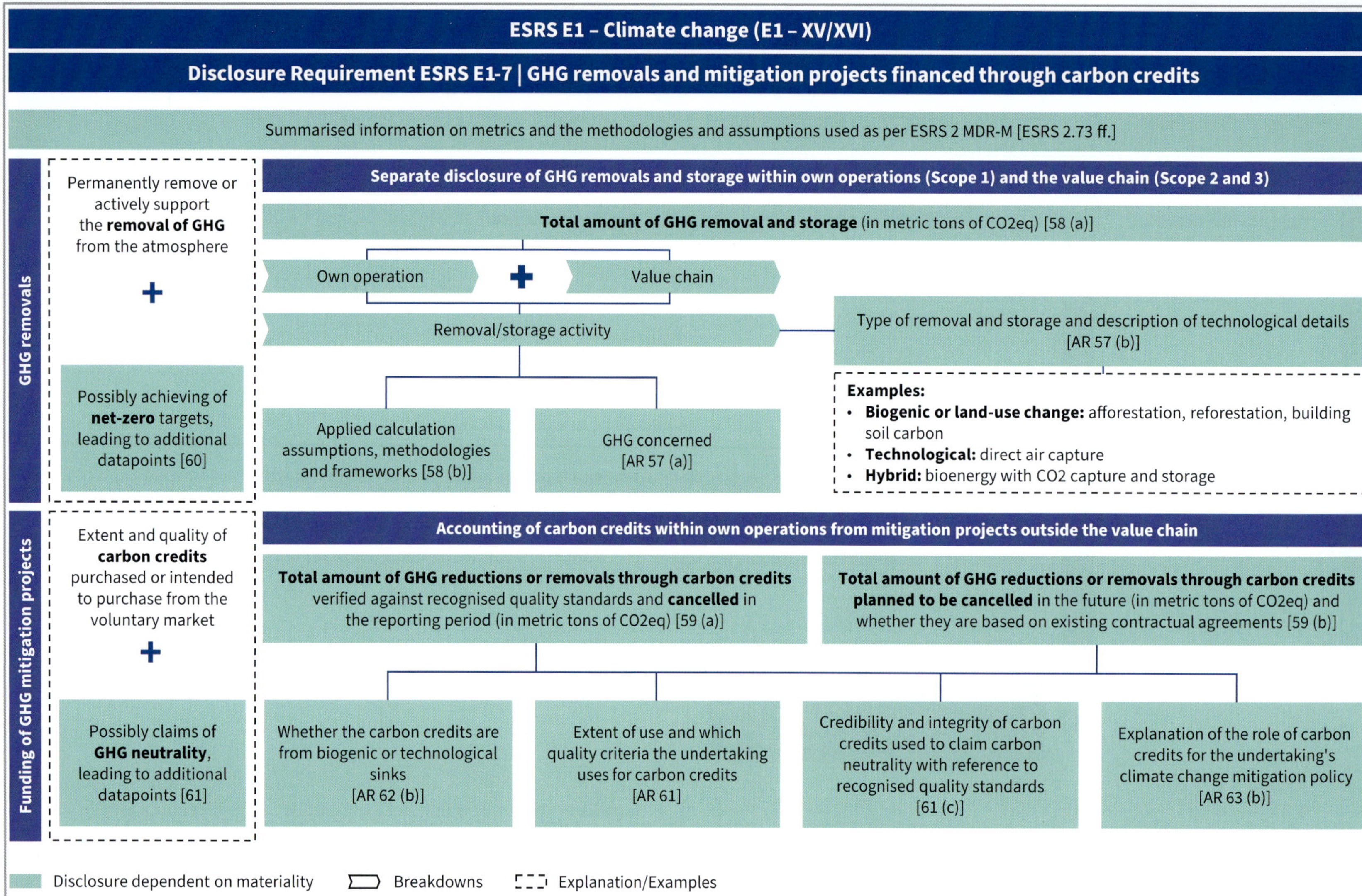
ESRS E1 – Climate change (E1 – XV/XVI)
Disclosure Requirement ESRS E1-7 | GHG removals and mitigation projects financed through carbon credits
Summarised information on metrics and the methodologies and assumptions used as per ESRS 2 MDR-M [ESRS 2.73 ff.]
GHG removals
Permanently remove or actively support the removal of GHG from the atmosphere
+
Possibly achieving of net-zero targets, leading to additional datapoints [60]
Separate disclosure of GHG removals and storage within own operations (Scope 1) and the value chain (Scope 2 and 3)
Total amount of GHG removal and storage (in metric tons of CO2eq) [58 (a)]
Own operation
+
Value chain
Removal/storage activity
Type of removal and storage and description of technological details [AR 57 (b)]
Examples:
• Biogenic or land-use change: afforestation, reforestation, building soil carbon
• Technological: direct air capture
• Hybrid: bioenergy with CO2 capture and storage
Applied calculation assumptions, methodologies and frameworks [58 (b)]
GHG concerned [AR 57 (a)]
Funding of GHG mitigation projects
Extent and quality of carbon credits purchased or intended to purchase from the voluntary market
+
Possibly claims of GHG neutrality, leading to additional datapoints [61]
Accounting of carbon credits within own operations from mitigation projects outside the value chain
Total amount of GHG reductions or removals through carbon credits verified against recognised quality standards and cancelled in the reporting period (in metric tons of CO2eq) [59 (a)]
Total amount of GHG reductions or removals through carbon credits planned to be cancelled in the future (in metric tons of CO2eq) and whether they are based on existing contractual agreements [59 (b)]
Whether the carbon credits are from biogenic or technological sinks [AR 62 (b)]
Extent of use and which quality criteria the undertaking uses for carbon credits [AR 61]
Credibility and integrity of carbon credits used to claim carbon neutrality with reference to recognised quality standards [61 (c)]
Explanation of the role of carbon credits for the undertaking's climate change mitigation policy [AR 63 (b)]
Disclosure dependent on materiality
Breakdowns
Explanation/Examples

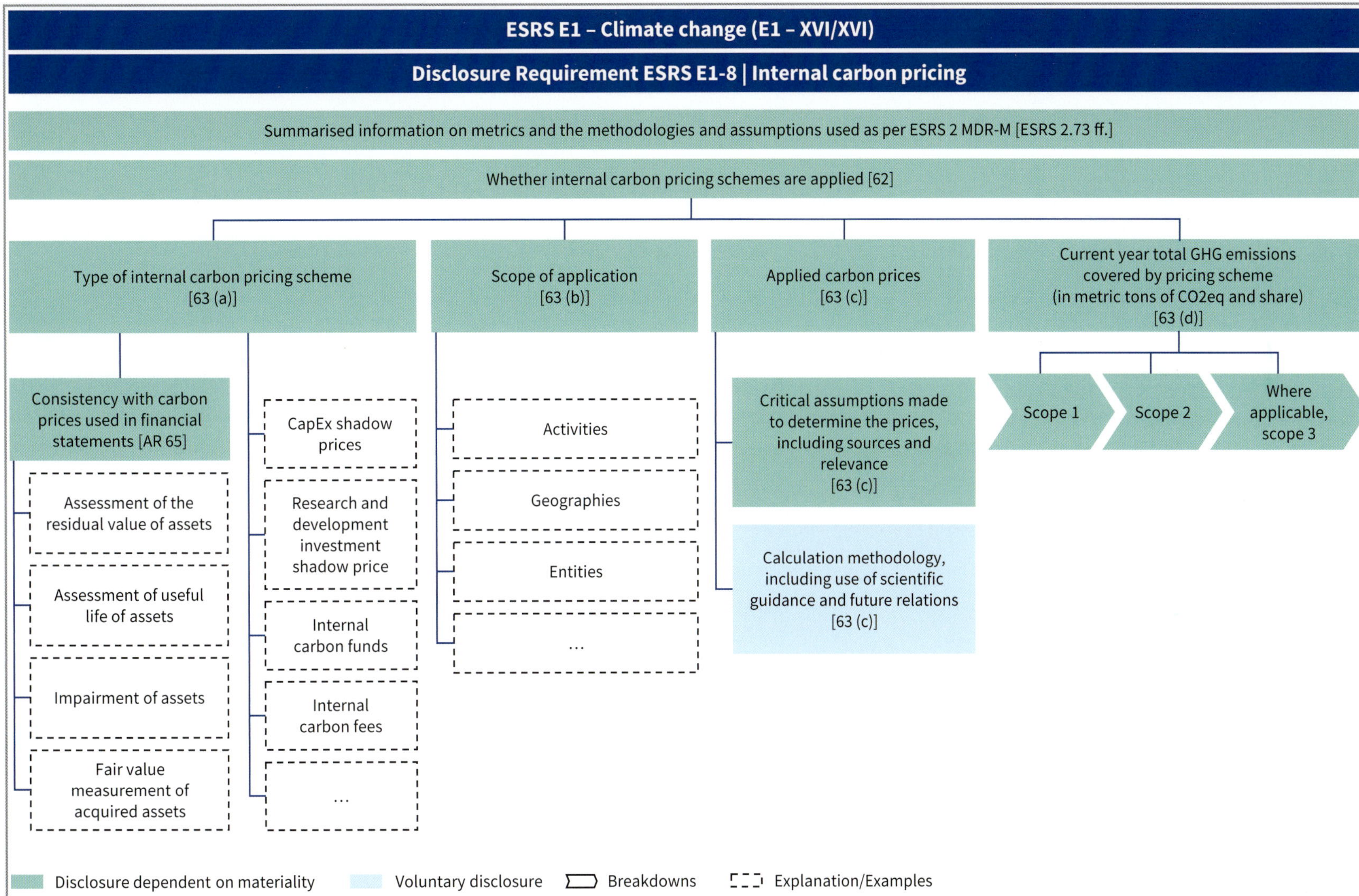
ESRS E1 – Climate change (E1 – XVI/XVI)
Disclosure Requirement ESRS E1-8 | Internal carbon pricing
Summarised information on metrics and the methodologies and assumptions used as per ESRS 2 MDR-M [ESRS 2.73 ff.]
Whether internal carbon pricing schemes are applied [62]
Type of internal carbon pricing scheme [63 (a)]
Scope of application [63 (b)]
Applied carbon prices [63 (c)]
Current year total GHG emissions covered by pricing scheme (in metric tons of CO2eq and share) [63 (d)]
Consistency with carbon prices used in financial statements [AR 65]
Assessment of the residual value of assets
Assessment of useful life of assets
Impairment of assets
Fair value measurement of acquired assets
CapEx shadow prices
Research and development investment shadow price
Internal carbon funds
Internal carbon fees
…
Activities
Geographies
Entities
…
Critical assumptions made to determine the prices, including sources and relevance [63 (c)]
Calculation methodology, including use of scientific guidance and future relations [63 (c)]
Scope 1
Scope 2
Where applicable, scope 3
Disclosure dependent on materiality
Voluntary disclosure
Breakdowns
Explanation/Examples

Intentionally left blank

ESRS E2 – Pollution

Objective:

Enabling understanding of:

- The material positive and negative actual and potential impacts of the undertaking on the pollution of air, water and soil.
- The plans and capacity of the undertaking to adapt its strategy and business model, in line with the transition to a sustainable economy and with the need to prevent, control and eliminate pollution in order to create a toxic-free environment with zero pollution.
- Any actions taken and the result of such actions to prevent or mitigate actual or potential negative impacts and to address risks and opportunities.
- The nature, type and extent of the undertaking's material risks and opportunities arising from its pollution-related impacts anddependencies, as well as the prevention, control, elimination or reduction of pollution (including where this results from regulations) and how it manages them.
- The financial effects on the undertaking over the short-, medium- and long-term of material risks and opportunities arising from its pollution-related impacts and dependencies.

Sustainability matters: Pollution of air, water, soil and living organisms and food resources, substances of concern, substances of very high concern, microplastics

Interaction with other ESRS:

- **Conjunctions**: ESRS 1 (General requirements), ESRS 2 (General disclosures)
- **Intersectional topics covered by other standards**: ESRS E1 (Greenhouse gases (carbon dioxide, methane, nitrous oxide, hydrofluorocarbons, perfluorocarbons, sulphur hexafluoride and nitrogen trifluoride) connected to air pollution)

  ESRS E5 (Pollution generated by waste)

  ESRS S3 (Material negative impacts on affected communities from pollution-related impacts)

Interaction with other frameworks:

- CLP-Regulation
- E-PRTR Regulation
- Industrial Emissions Directive
- REACH-Regulation

Overall evaluation:

The overall evaluation contains a subjective assessment of the corresponding standard as narrative or quantitative and a count of all disclosures included in the corresponding standard. The number of disclosures was determined using professional judgement and depends on various factors. Due to the subjectivity involved, other counts may deviate.

ESRS E2 – Pollution (E2 – I/VII)

Structure and content

General disclosures

ESRS 2 XVI/XXII

ESRS 2 IRO-1

Description of the **processes** to **identify** and **assess** material pollution-related **IROs**

Information on whether and how site locations and activities have been screened including the methodologies, assumptions and tools used and whether and how consultations were conducted, in particular, with affected communities. The LEAP approach may be considered.

Impact, risk and opportunity management

ESRS 2 MDR-P

E2-1

Policies related to pollution

Information on the extent to which policies address the identification, assessment, management and/or remediation of material pollution-related IROs.

ESRS 2 MDR-A

E2-2

Actions and **resources** related to pollution

Information on key actions taken and planned and the resources allocated to their implementation to achieve pollution-related policy objectives and targets.

Metrics and targets

ESRS 2 MDR-T

E2-3

Targets related to pollution

Information on targets set to support pollution-related policies and address material pollution-related IROs.

ESRS 2 MDR-M | V/VII

E2-4

Pollution of air, water and soil

Information on pollutants emitted and microplastics generated or used during own production processes or procured and that leave the facilities as emissions, as products, or as part of products or services.

ESRS 2 MDR-M | VI/VII

E2-5

Substances of concern and substances of very high concern

Information on the production, use, distribution, commercialisation and import/export of substances of concern and substances of very high concern, on their own, in mixtures or in articles.

ESRS 2 MDR-M | VII/VII

E2-6

Anticipated financial effects from material pollution-related risks and opportunities

Information on anticipated financial effects due to material risks arising from pollution-related impacts and dependencies.

Information on anticipated financial effects due to material opportunities related to pollution prevention and control.

ESRS E2 – Pollution (E2 – II/VII)

Key disclosures concerning IRO management

ESRS 2 MDR-A

Actions and resources related to pollution [E2-2]

- List of key actions taken and planned, their expected outcomes and their contribution to the achievement of policy objectives and targets [ESRS 2 MDR-A, 68 (a)]
- Key actions taken (along with results) to provide remedy for those harmed by actual material impacts [ESRS 2 MDR-A, 68 (d)]
- Whether the ability to implement each key action depends on specific preconditions [ESRS 2 MDR-A, 69 (a)]
- Amount of current and future financial resources for the implementation of a key action or action plan [ESRS 2 MDR-A, 69 (b), (c)]
- To which layer in the mitigation hierarchy actions or resources can be allocated [19]

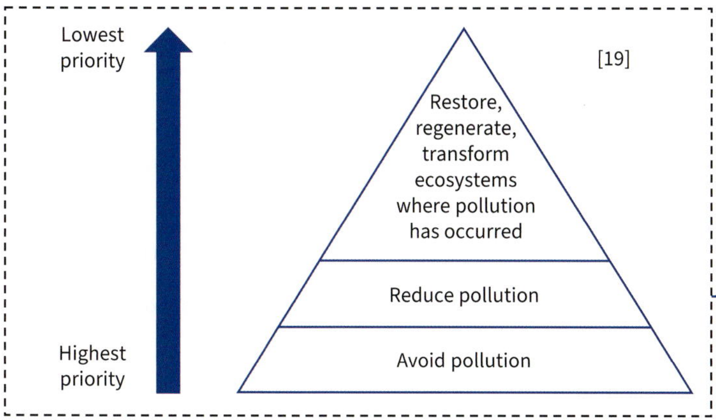

Disclosure dependent on materiality | Voluntary disclosure | Explanation/Examples

ESRS E2 – Pollution (E2 – III/VII)

Key disclosures concerning metrics and targets – 1/2

Targets related to pollution [E2-3] (ESRS 2 MDR-T)

- Performance against each disclosed target, including an analysis of trends or significant changes in the performance [ESRS 2 MDR-T, 80 (j)]
- Information on how each target is monitored and reviewed and whether progress is in line with what had been initially planned [ESRS 2 MDR-T, 80 (j)]

Pollution of air, water and soil [E2-4] (ESRS 2 MDR-M) (V/VII)

- Whether the measurement of the metric is validated by an external body other than the assurance provider and, if so, which body [ESRS 2 MDR-M, 77 (b)]
- Amounts of each pollutant listed in Annex 2 of E-PRTR Regulation emitted to: [28 (a)]
  - Air (except for GHG emissions),
  - Water and
  - Soil
- Amount of microplastics generated and amount of microplastic used [28 (b)]
- Changes over time in pollutants emitted and microplastics generated or used [30 (a)]
- Disclosures related to the Industrial Emissions Directive [AR 25]

Breakdowns:
- At site level [AR 22]
- Type of source [AR 22]
- Sector [AR 22]
- Geographical area [AR 22]

Disclosure dependent on materiality | Voluntary disclosure | Breakdowns

ESRS E2 – Pollution (E2 – IV/VII)

Key disclosures concerning metrics and targets– 2/2

Substances of concern and substances of very high concern [E2-5]

ESRS 2 MDR-M — VI/VII

- Whether the measurement of the metric is validated by an external body other than the assurance provider and, if so, which body [ESRS 2 MDR-M, 77 (b)]
- Total amount of substances of concern that are generated or used during the production or that are procured [34]
- Total amount of substances of very high concern that are generated or used during the production or that are procured [35]
  - Main hazard classes as defined in Annex II
- Total amount of substances of concern that leave the facilities as emissions, as products, or as part of products or services [34]
- Total amount of substances of very high concern that leave the facilities as emissions, as products, or as part of products or services [35]
  - as emissions [AR 22]
  - as products [AR 22]
  - as part of products [AR 22]
  - as services [AR 22]
    - Main hazard classes as defined in Annex II

Anticipated financial effects [E2-6]

ESRS 2 MDR-M — VII/VII

- Whether the measurement of the metric is validated by an external body other than the assurance provider and, if so, which body [ESRS 2 MDR-M, 77 (b)]
- Anticipated financial effects in monetary terms due to material risks before considering pollution-related actions [39 (a)]
- Anticipated financial effects in monetary terms due to material opportunities before considering pollution-related actions [39 (a)]
- Share of net revenue made with products and services that are or that contain substances of concern and share of net revenue made with products and services that are or that contain substances of very high concern [40 (a)]
- Operating expenditures and capital expenditures incurred in the reporting period in conjunction with major incidents and deposits [40 (b)]
- Provisions for environmental protection and remediation costs [40 (c)]
- Assessment of related products and services at risk over the short-, medium- and long-term, explaining how these are defined, how financial amounts are estimated and which critical assumptions are made [AR 33]

Disclosure dependent on materiality | Disclosure dependent on materiality with phase-in | Voluntary disclosure | Breakdowns

ESRS E2 – Pollution (E2 – V/VII)

Disclosure Requirement ESRS E2-4 | Pollution of air, water and soil

Summarised information on metrics and the methodologies and assumptions used as per ESRS 2 MDR-M [ESRS 2.73 ff.]

Disclosure of consolidated amounts including emissions from facilities financially and operationally controlled.
Include only emissions from facilities exceeding the applicable threshold value specified in Annex II of the E-PRTR Regulation. [29]

Pollution of air, water and soil emitted through own operations [26] (except for GHG emissions)

Amounts of each pollutant listed in Annex II of the E-PRTR Regulation emitted to: [28 (a)]

Air + Water + Soil

Generation and use of microplastics [26]

Amounts of microplastics… [28 (b)]

…that are generated [28 (b)] + …used during production processes [28 (b)] + …that are procured [AR 20] +

…that leave the facilities as emissions [AR 20] + …that leave the facilities as products [AR 20] + …that leave the facilities as part of products/services [AR 20]

Concerning estimates: standard, sectoral study or sources forming the basis, the possible degree of uncertainty and the range of estimates reflecting the measurement uncertainty [31]

Reference to information reported under other existing legislation (e.g. Industrial Emission Directive, E-PRTR Regulation, etc.) [AR 24]

Contextual information: [30]

- Description of changes over time regarding the quantitative metrics
- Description of measurement technologies
- Description of the process(es) to collect data, including the type of data needed and the information sources

At site level [AR 22] | Sector [AR 22]

Type of source [AR 22] | Geographical area [AR 22]

Disclosure dependent on materiality | Voluntary disclosure | Breakdowns | Explanation/Examples

ESRS E2 – Pollution (E2 – VI/VII)

Disclosure Requirement ESRS E2-5 | Substances of concern and substances of very high concern

Summarised information on metrics and the methodologies and assumptions used as per ESRS 2 MDR-M [ESRS 2.73 ff.]

The disclosures shall include substances in the undertaking's own operations and those procured (e.g. embedded in ingredients, semi-finished products, or the final product). [AR 28]

Total amounts of substances of concern… [34]

Total amounts of substances of very high concern… [35]

…that are generated or used during the production or that are procured

- Main hazard classes as defined in Annex II [34, 35]

…that leave the facilities as emissions, products, part of products or services

- as emissions [AR 22]
- as products [AR 22]
- as part of products [AR 22]
- as services [AR 22]
- Main hazard classes as defined in Annex II [34, 35]

For example, substances that are classified in Part 3 of Annex VI of the CLP-Regulation in one of the following hazard classes or hazard categories:

- Carcinogenicity categories 1 and 2 **or**
- Germ cell mutagenicity categories 1 and 2 **or**
- Reproductive toxicity categories 1 and 2
- Respiratory sensitisation category 1 **or**
- Skin sensitisation category 1 **or**
- Skin sensitisation category 1

Substances of very high concern are listed in the European Chemical Agency Candidate List which is a subsequent document of the REACH-Regulation

Reference to information reported under other existing legislation (e.g. Industrial Emission Directive, E-PRTR Regulation, etc.) [AR 30]

Disclosure dependent on materiality | Voluntary disclosure | Breakdowns | Explanation/Examples

ESRS E2 – Pollution (E2 – VII/VII)

Disclosure Requirement ESRS E2-6 | Anticipated financial effects

DR requests:

- Summarised information on metrics and the methodologies and assumptions used as per ESRS 2 MDR-M [ESRS 2.73 ff.]
- Anticipated financial effects in monetary terms due to material risks before considering pollution-related actions [39 (a)]
  - Scope of the disclosure: risks that have or could reasonably be expected to have a material influence on the undertaking's: [38 (a)]
    - Financial position
    - Financial performance
    - Cash flows
      - Short-term
      - Medium-term
      - Long-term
  - Quantification of the anticipated financial effects due to material risks may be a single amount or a range. [AR 34]
- Anticipated financial effects in monetary terms due to material opportunities before considering pollution-related actions [39 (a)]

DR shall include:

- Qualitative information on the anticipated financial effects before considering pollution-related actions, if a quantification is not possible without undue cost or effort [39 (a)]
  - For financial effects arising from opportunities, a quantification is not required if it would result in disclosure that does not meet the qualitative characteristics of information. [39 (a)] (ESRS 1 VI/XXIX)
- Description of the effects considered, impacts and dependencies to which they relate and time horizons in which those effects are likely to materialise [39 (b)] + Critical assumptions used to quantify the anticipated financial effects, as well as sources and level of uncertainty of those assumptions [39 (c)]
- Quantification of specific financial effects in form of revenue, operating and capital expenditures and provisions and contextual information [40 (a), (b), (c), 41]

Assessment: [AR 33]

- Assessment of the undertaking's related products and services at risk
  - Short-term
  - Medium-term
  - Long-term
- Definition of the undertaking's products and services
- How financial amounts are estimated
- Which critical assumptions are made

Legend: Disclosure dependent on materiality with phase-in | Voluntary disclosure | Explanation/Examples (dashed)

ESRS E3 – Water and marine resources

Objective:

Enabling understanding of:

- The material positive and negative actual and potential impacts of the undertaking on water and marine resources.
- The plans and capacity of the undertaking to adapt its strategy and business model in line with the promotion of sustainable water use based on long-term protection of available water resources, protection of aquatic ecosystems and restoration of freshwater and marine habitats.
- Any actions taken and the result of such actions to prevent or mitigate material actual or potential negative impacts, to protect water and marine resources, also with reference to reduction of water consumption, and to address risks and opportunities.
- Whether, how and to what extent the undertaking contributes to the European Green Deal's ambitions for fresh air, clean water, healthy soil and biodiversity, as well as to the sustainability of the blue economy and fisheries sectors.
- The nature, type and extent of the undertaking's material risks and opportunities arising from its impacts and dependencies on water and marine resources and how it manages them.
- The financial effects on the undertaking over the short-, medium- and long-term of material risks and opportunities arising from its impacts and dependencies on water and marine resources.

Sustainability matters:

Water (water consumption, water withdrawals, water discharges, water discharges in the oceans), marine resources (extraction and use of marine resources)

Interaction with other ESRS:

- **Conjunctions**

  ESRS 1 (General requirements), ESRS 2 (General disclosures)

- **Intersectional topics covered by other standards**

  ESRS E1 (Acute and chronic physical risks from water and ocean-related hazards caused or exacerbated by climate change)
  ESRS E2 (Emissions to water and the oceans and the use and generation of microplastics)
  ESRS E4 (Conservation and sustainable use of and impact on freshwater aquatic ecosystems as well as the oceans and seas)
  ESRS E5 (Waste management, including plastic and wastewater)
  ESRS S3 (Material negative impacts on affected communities from water and marine resource-related impacts)

Interaction with other frameworks:

- Alliance for Water Stewardship
- Environmental Action Programme
- Marine Strategy Framework Directive
- Maritime Spatial Planning Directive
- Science-Based Targets Initiative for Nature
- SDGs (6 Clean Water and Sanitation and 14 Life Below Water)
- Water Framework Directive

Overall evaluation:

The overall evaluation contains a subjective assessment of the corresponding standard as narrative or quantitative and a count of all disclosures included in the corresponding standard. The number of disclosures was determined using professional judgement and depends on various factors. Due to the subjectivity involved, other counts may deviate.

ESRS E3 – Water and marine resources (E3 – I/X)

Structure and content

General disclosures

ESRS 2 XVI/XXII

ESRS 2 IRO-1

Description of the **processes** to **identify** and **assess** material water and marine resources-related **IROs**

Information on whether and how assets and activities have been screened including the methodologies, assumptions and tools used and whether and how consultations were conducted, in particular, with affected communities. The LEAP approach may be considered.

Impact, risk and opportunity management

ESRS 2 MDR-P | VI/X

E3-1

Policies related to water and marine resources

Information on the extent to which policies address the identification, assessment, management and/or remediation of material water and marine resources-related IROs.

ESRS 2 MDR-A | VII/X

E3-2

Actions and **resources** related to water and marine resources

Information on key actions taken and planned and the resources allocated to their implementation to achieve water and marine resources-related policy objectives and targets.

Metrics and targets

ESRS 2 MDR-T | VIII/X

E3-3

Targets related to water and marine resources

Information on targets set to support water and marine resources-related policies and address material water and marine resources-related IROs.

ESRS 2 MDR-M | IX/X

E3-4

Water consumption

Information on the undertaking's water consumption and any progress in relation to the targets, including water consumption, water consumption in areas at water risk and high water stress, recycled and stored water as well as the water intensity (water consumption per net revenue).

ESRS 2 MDR-M | X/X

E3-5

Anticipated financial effects from material water and marine resources-related risks and opportunities

Information on anticipated financial effects due to material risks arising from water and marine resources-related impacts and dependencies.

Information on anticipated financial effects due to material opportunities related to water and marine resources.

ESRS E3 – Water and marine resources (E3 – II/X)

Key disclosures concerning IRO management

ESRS 2 MDR-A

Actions and resources related to water and marine resources [E3-2]

VII/X

- List of key actions taken and planned, their expected outcomes and their contribution to the achievement of policy objectives and targets [ESRS 2 MDR-A, 68 (a)]
- Key actions taken (along with results) to provide remedy for those harmed by actual material impacts [ESRS 2 MDR-A, 68 (d)]
- Whether the ability to implement each key action depends on specific preconditions [ESRS 2 MDR-A, 69 (a)]
- Amount of current and future financial resources for the implementation of a key action or action plan [ESRS 2 MDR-A, 69 (b), (c)]
- Specification to which layer in the mitigation hierarchy an action and resources can be allocated [18]

Disclosure dependent on materiality | Voluntary disclosure

ESRS E3 – Water and marine resources (E3 – III/X)

Key disclosures concerning metrics and targets – 1/3

ESRS 2 MDR-T

Targets related to water and marine resources [E3-3]

VIII/X

- Performance against each disclosed target, including an analysis of trends or significant changes in the performance [ESRS 2 MDR-T, 80 (j)]
- Information on how each target is monitored and reviewed and whether progress is in line with what had been initially planned [ESRS 2 MDR-T, 80 (j)]

Disclosure dependent on materiality

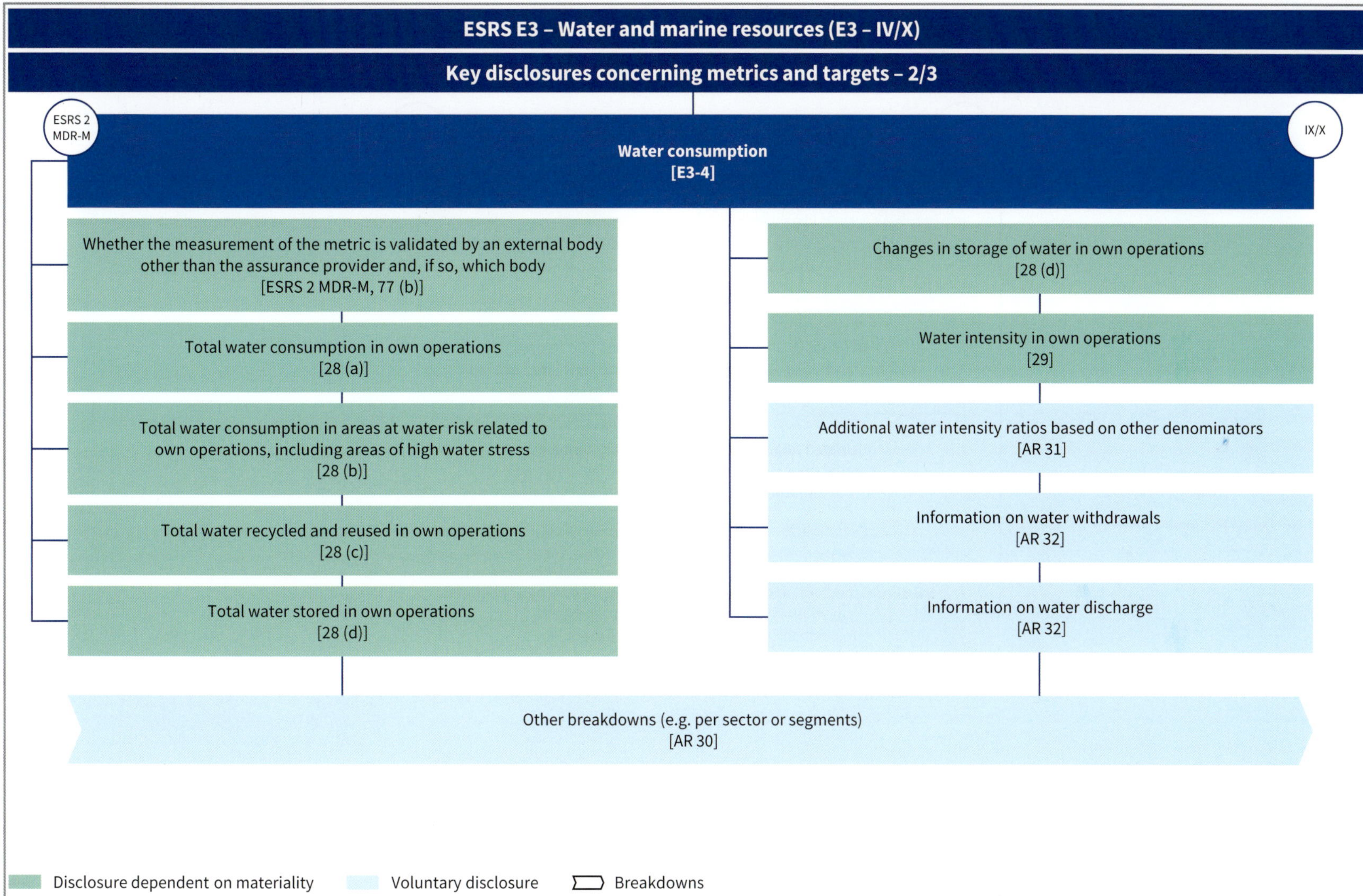
ESRS E3 – Water and marine resources (E3 – IV/X)
Key disclosures concerning metrics and targets – 2/3
ESRS 2 MDR-M
IX/X
Water consumption [E3-4]
Whether the measurement of the metric is validated by an external body other than the assurance provider and, if so, which body [ESRS 2 MDR-M, 77 (b)]
Total water consumption in own operations [28 (a)]
Total water consumption in areas at water risk related to own operations, including areas of high water stress [28 (b)]
Total water recycled and reused in own operations [28 (c)]
Total water stored in own operations [28 (d)]
Changes in storage of water in own operations [28 (d)]
Water intensity in own operations [29]
Additional water intensity ratios based on other denominators [AR 31]
Information on water withdrawals [AR 32]
Information on water discharge [AR 32]
Other breakdowns (e.g. per sector or segments) [AR 30]
Disclosure dependent on materiality
Voluntary disclosure
Breakdowns

ESRS E3 – Water and marine resources (E3 – V/X)

Key disclosures concerning metrics and targets – 3/3

ESRS 2 MDR-M

Anticipated financial effects [E3-5]

X/X

- Whether the measurement of the metric is validated by an external body other than the assurance provider and, if so, which body [ESRS 2 MDR-M, 77 (b)]
- Anticipated financial effects in monetary terms due to material risks before considering water and marine resources-related actions [33 (a)]
- Anticipated financial effects in monetary terms due to material opportunities before considering water and marine resources-related actions [33 (a)]
- Assessment of related products and services at risk over the short-, medium- and long-term, explaining how these are defined, how financial amounts are estimated and which critical assumptions are made [AR 33]

Disclosure dependent on materiality with phase-in | Voluntary disclosure

ESRS E3 – Water and marine resources (E3 – VI/X)

Disclosure Requirement ESRS E3-1 | Policies related to water and marine resources

Summarised information on the key content, scope and communication of the policies as per ESRS 2 MDR-P [ESRS 2.63 ff.]

Sites located in an area of high water stress: statement on lack of policies and reasons for not having adopted such policies [13]

Sites located in an area of high water stress: timeframe in which the undertaking aims to adopt such a policy [13]

Whether and how policies address:

- Water management, including: [12 (a)]
  - Use and sourcing of water and marine resources in own operations
  - Water treatment
  - Prevention and abatement of water pollution
- Product and service design in view of addressing water-related issues and the preservation of marine resources [12 (b)]
- Commitment to reduce material water consumption in areas at water risk in own operations and along the value chain [12 (c)]

Information on policies which:

- Contribute to good ecological and chemical quality of surface water bodies and good chemical quality and quantity of groundwater bodies [AR 18 (a)]
- Minimise material impacts and risks and implement mitigation measures that aim to maintain e.g. the value and function of priority services [AR 18 (b)]
- Avoid impacts on affected communities [AR 18 (c)]

Whether policies:

- Address sustainable oceans and seas [14]
- Prevent further deterioration and protect and enhance the status of water bodies and aquatic ecosystems [AR 17 (a)]
- Promote sustainable water use based on a long-term protection of available water resources [AR 17 (b)]
- Aim at enhanced protection and improvement of the aquatic environment [AR 17 (c)]
- Promote a good environmental status of marine water [AR 17 (d)]
- Promote reduction of water withdrawals and water discharges [AR 17 (e)]

Disclosure dependent on materiality | Voluntary disclosure

ESRS E3 – Water and marine resources (E3 – VII/X)

Disclosure Requirement ESRS E3-2 | Actions and resources related to water and marine resources

Summarised information on actions and resources to manage IROs as per ESRS 2 MDR-A [ESRS 2.66 ff.]

Specification to which layer in the mitigation hierarchy an action and resources can be allocated [18]

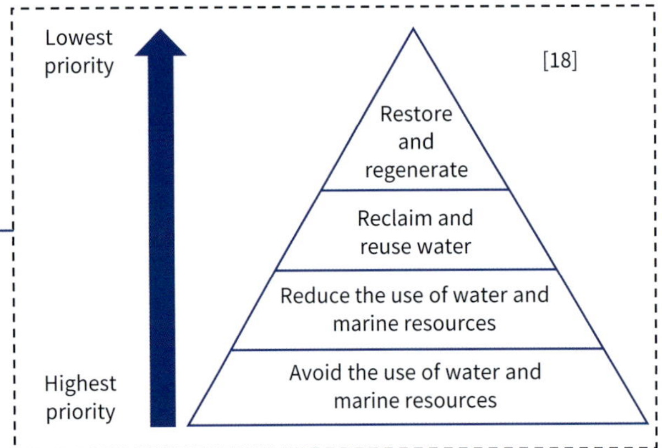

Specification of actions and resources in relation to areas at water risk, including areas of high water stress [19]

Information on specific collective actions or action plans, including information on other parties, the project, the contribution of the undertaking to the collective project and its sponsors and other participants [AR 20]

+

Alliance for Water Stewardship provides useful guidance. [AR 19]

Disclosure dependent on materiality | Voluntary disclosure | Explanation/Examples

ESRS E3 – Water and marine resources (E3 – VIII/X)

Disclosure Requirement ESRS E3-3 | Targets related to water and marine resources

Summarised information on targets and the performance against those as per ESRS 2 MDR-T [ESRS 2.78 ff.]

Whether targets are mandatory (required by legislation) or voluntary [25]

Details on ecological thresholds [24]

Whether the target addresses shortcomings related to the Substantial Contribution criteria for Water and Marine Resources and whether the target addresses shortcomings related to the Do No Significant Harm criteria [AR 27]

Whether and how targets relate to:

- Management of material IROs related to areas at water risk, including improvement of the water quality [23 (a)]
- Responsible management of IROs related to marine resources, including the nature and quantity of marine resources-related commodities [23 (b)]

Further considerations:

- In case of reference to ecological thresholds when setting targets: The undertaking may refer to guidance provided by the Science-Based Targets Initiative for Nature in its interim guidance or any other guidance with a scientifically acknowledged methodology. [AR 22]
- Targets may cover the undertaking's own operations and/or its value chain. [AR 26]

Whether and how targets relate to the reduction of:

- Water consumption, including an explanation of how those targets relate to areas at water risk, including areas of high water stress [23 (c)]
- Water withdrawals, including from polluted soils and aquifers, and water withdrawn and treated for remediation purposes [AR 23 (a), AR 24]
- Water discharges, including to groundwater such as reinjection to aquifers, or water returning to a groundwater source via a soakaway or a swale [AR 23 (b), AR 25]

Disclosure dependent on materiality | Voluntary disclosure | Explanation/Examples

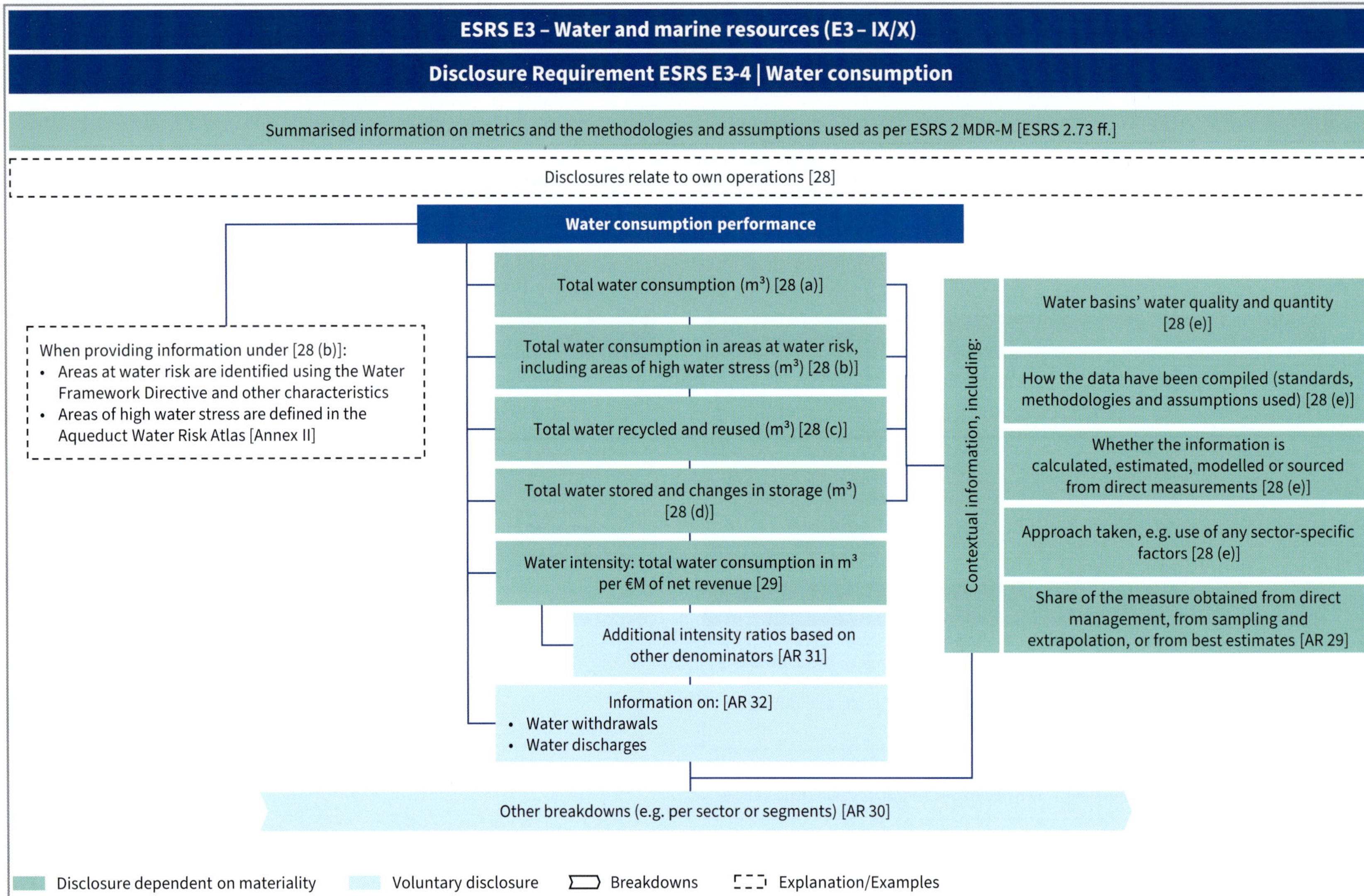
ESRS E3 – Water and marine resources (E3 – IX/X)
Disclosure Requirement ESRS E3-4 | Water consumption
Summarised information on metrics and the methodologies and assumptions used as per ESRS 2 MDR-M [ESRS 2.73 ff.]
Disclosures relate to own operations [28]
Water consumption performance
When providing information under [28 (b)]:
• Areas at water risk are identified using the Water Framework Directive and other characteristics
• Areas of high water stress are defined in the Aqueduct Water Risk Atlas [Annex II]
Total water consumption (m³) [28 (a)]
Total water consumption in areas at water risk, including areas of high water stress (m³) [28 (b)]
Total water recycled and reused (m³) [28 (c)]
Total water stored and changes in storage (m³) [28 (d)]
Water intensity: total water consumption in m³ per €M of net revenue [29]
Additional intensity ratios based on other denominators [AR 31]
Information on: [AR 32]
• Water withdrawals
• Water discharges
Contextual information, including:
Water basins' water quality and quantity [28 (e)]
How the data have been compiled (standards, methodologies and assumptions used) [28 (e)]
Whether the information is calculated, estimated, modelled or sourced from direct measurements [28 (e)]
Approach taken, e.g. use of any sector-specific factors [28 (e)]
Share of the measure obtained from direct management, from sampling and extrapolation, or from best estimates [AR 29]
Other breakdowns (e.g. per sector or segments) [AR 30]
Disclosure dependent on materiality
Voluntary disclosure
Breakdowns
Explanation/Examples

ESRS E3 – Water and marine resources (E3 – X/X)

Disclosure Requirement ESRS E3-5 | Anticipated financial effects

DR requests:

- Summarised information on metrics and the methodologies and assumptions used as per ESRS 2 MDR-M [ESRS 2.73 ff.]
- Anticipated financial effects in monetary terms due to material risks before considering water and marine resources-related actions [33 (a)]
  - Scope of the disclosure: risks that have or could reasonably be expected to have a material influence on the undertaking's: [32 (a)]
    - Financial position
    - Financial performance
    - Cash flows
      - Short-term
      - Medium-term
      - Long-term
  - Quantification of the anticipated financial effects due to material risks may be a single amount or a range. [AR 34]
- Anticipated financial effects in monetary terms due to material opportunities before considering water and marine resources-related actions [33 (a)]

DR shall include:

- Qualitative information on the anticipated financial effects before considering water and marine resources-related actions, if a quantification is not possible without undue cost or effort [33 (a)]
  - For financial effects arising from opportunities, a quantification is not required if it would result in disclosure that does not meet the qualitative characteristics of information. [33 (a)] (ESRS 1 VI/XXIX)
- Description of the effects considered, impacts and dependencies to which they relate and time horizons in which those effects are likely to materialise [33 (b)]
  \+ Critical assumptions used to quantify the anticipated financial effects, as well as sources and level of uncertainty of those assumptions [33 (c)]

Assessment: [AR 33]

- Assessment of the undertaking's related products and services at risk
  - Short-term
  - Medium-term
  - Long-term
- Definition of the undertaking's products and services
- How financial amounts are estimated
- Which critical assumptions are made

Legend: Disclosure dependent on materiality with phase-in | Voluntary disclosure | Explanation/Examples

Intentionally left blank

ESRS E4 – Biodiversity and ecosystems

Objective:

Enabling understanding of:

- The material positive and negative actual and potential impacts of the undertaking on biodiversity and ecosystems, including the extent to which it contributes to the drivers of biodiversity and ecosystem loss and degradation.
- The plans and capacity of the undertaking to adapt its strategy and business model in line with EU directives and other frameworks.
- Any actions taken and result of such actions to prevent or mitigate actual or potential negative impacts, to protect and restore biodiversity and ecosystems and to address risks and opportunities.
- The nature, type and extent of the undertaking's material risks, dependencies and opportunities arising from biodiversity and ecosystems and how it manages them.
- The financial effects on the undertaking over the short-, medium- and long-term of material risks and opportunities arising from its impacts and dependencies on biodiversity and ecosystems.

Sustainability matters:

Direct impact drivers of biodiversity loss (climate change, land-use change, freshwater-use change, sea-use change, direct exploitation, invasive alien species, pollution, others), impacts on the state of species (e.g. population size, global extinction risk), impacts on extent and condition of ecosystems (e.g. land degradation, desertification, soil sealing), impacts and dependencies on ecosystem services

Interaction with other ESRS:

- **Conjunctions**

ESRS 1 (General requirements), ESRS 2 (General disclosures)

- **Intersectional topics covered by other standards**

ESRS E1 (Climate change as a direct impact driver of biodiversity and ecosystems change), ESRS E2 (Pollution as a direct impact driver of biodiversity and ecosystems change), ESRS S3 (Material negative impacts on affected communities from impacts on biodiversity and ecosystems)

Interaction with other frameworks:

- Birds and Habitats Directives
- Common International Classification of Ecosystem Services
- Convention for Biological Diversity
- Delegated Regulation supplementing the EU Taxonomy by establishing technical screening criteria
- EU Biodiversity Strategy for 2030
- IUCN Red List of Threatened Species
- Kunming-Montreal Global Biodiversity Framework
- Marine Strategy Framework Directive
- Natura 2000 network of protected areas
- Planetary Boundaries
- Platform on Biodiversity and Ecosystem Services
- Specific indicators developed by the Joint Research Center
- UNESCO World Heritage Sites

Applicability:

Undertakings or groups not exceeding the average number of 750 employees may omit the information specified in the DRs of ESRS E4 for the first two years of preparation of their sustainability statement. In this case the undertaking discloses the requirements according to ESRS 2.17.

ESRS 2 X/XXII

Overall evaluation:

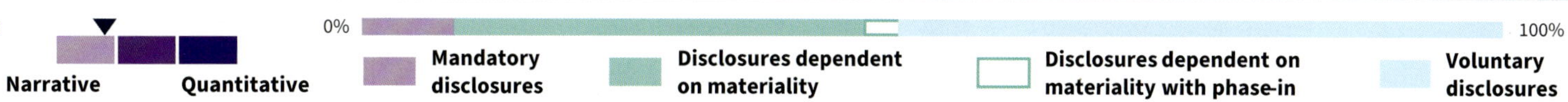

The overall evaluation contains a subjective assessment of the corresponding standard as narrative or quantitative and a count of all disclosures included in the corresponding standard. The number of disclosures was determined using professional judgement and depends on various factors. Due to the subjectivity involved, other counts may deviate.

ESRS E4 – Biodiversity and ecosystems (E4 – I/XI)

Structure and content

General disclosures

E4-1 (VII/XI)

Transition plan and consideration of biodiversity and ecosystems in strategy and business model

Information on the resilience of the strategy and business model in relation to biodiversity and ecosystems and on the compatibility of the strategy and business model with regard to relevant local, national and global public policy targets related to biodiversity and ecosystems.

ESRS 2 SBM-3

Material IROs and their interaction with **strategy and business model**

Categorisation of material sites in own operations, whether material negative impacts with regards to land degradation, desertification or soil sealing exist and whether threatened species are affected.

ESRS 2 IRO-1 (ESRS 2 XVI/XXII)

Description of the **processes** to **identify** and **assess** biodiversity and ecosystem-related **IROs**

Information on e.g. whether sites exist from which impacts occur and/or which are located in or near biodiversity-sensitive areas, whether the activities negatively affect these and whether and how consultations were conducted, in particular, with affected communities. The LEAP approach may be considered.

Impact, risk and opportunity management

E4-2 (ESRS 2 MDR-P)

Policies related to biodiversity and ecosystems

Information on the extent to which policies address the identification, assessment, management and/or remediation of material biodiversity and ecosystem-related IROs.

E4-3 (ESRS 2 MDR-A)

Actions and **resources** related to biodiversity and ecosystems

Information on key actions taken and planned and the resources allocated to their implementation to achieve biodiversity and ecosystem-related policy objectives and targets.

Metrics and targets

E4-4 (ESRS 2 MDR-T)

Targets related to biodiversity and ecosystems

Information on targets set to support biodiversity and ecosystem-related policies and address material biodiversity and ecosystem-related IROs.

E4-6 (ESRS 2 MDR-M) (XI/XI)

Anticipated financial effects from material biodiversity and ecosystem-related risks and opportunities

Information on anticipated financial effects due to material risks arising from biodiversity and ecosystem-related impacts and dependencies.

Information on anticipated financial effects due to material opportunities related to biodiversity and ecosystems.

E4-5 (ESRS 2 MDR-M) (IX/XI)

Impact metrics related to biodiversity and ecosystems change

Performance indicators of the undertaking against material impacts with regards to e.g.:

- Sites in or near biodiversity-sensitive areas
- Impact drivers of land-use change, freshwater-use change and/or sea-use change
- Introduction of invasive alien species
- State of species
- Ecosystems

ESRS E4 – Biodiversity and ecosystems (E4 – II/XI)

Key disclosures concerning general disclosures

Transition plan and consideration of biodiversity and ecosystems in strategy and business model [E4-1]

VII/XI

- Description of the results of the resilience analysis of the strategy and business model in relation to biodiversity and ecosystems [13 (e)]
- How the undertaking contributes to addressing biodiversity and ecosystem impact drivers and its possible mitigation actions following the mitigation hierarchy and the main path-dependencies and locked-in assets and resources that are associated with biodiversity and ecosystems change [AR 1 (d)]
- Quantification of the investments and funding supporting the implementation of the transition plan, with a reference to the key performance indicators of taxonomy-aligned CapEx and where relevant the CapEx plans [AR 1 (e)]
- How biodiversity offsets are used as part of the transition plan and if so, where the offsets are planned to be used, the extent of the use in relation to the overall transition plan and whether the mitigation hierarchy was considered [AR 1 (g)]

Material IROs and their interaction with strategy and business model [ESRS 2 SBM-3]

- Whether material negative impacts with regards to land degradation, desertification or soil sealing have been identified [16 (b)]
- Whether the undertaking has operations that affect threatened species [16 (c)]

Description of processes to identify and assess material biodiversity and ecosystem-related IROs [ESRS 2 IRO-1]

- Whether and how the undertaking identified and assessed
  - Actual and potential impacts
  - Dependencies
  - Transition and physical risks and opportunities on biodiversity and ecosystems
  - At own site locations and
  - In the value chain

  [17]
- Whether activities related to sites located in or near biodiversity-sensitive areas negatively affect these areas [19 (a)]

Mandatory disclosure | Disclosure dependent on materiality | Voluntary disclosure

ESRS E4 – Biodiversity and ecosystems (E4 – III/XI)

Key disclosures concerning IRO management

ESRS 2 MDR-P — Policies related to biodiversity and ecosystems [E4-2]

- Connections and alignment with other global goals and agreements [AR 13]

ESRS 2 MDR-A — Actions and resources related to biodiversity and ecosystems [E4-3]

- List of key actions taken and planned, their expected outcomes and their contribution to the achievement of policy objectives and targets [ESRS 2 MDR-A, 68 (a)]
- Key actions taken (along with results) to provide remedy for those harmed by actual material impacts [ESRS 2 MDR-A, 68 (d)]
- Whether the ability to implement each key action depends on specific preconditions [ESRS 2 MDR-A, 69 (a)]
- Amount of current and future financial resources for the implementation of a key action or action plan [ESRS 2 MDR-A, 69 (b), (c)]
- Whether the undertaking used biodiversity offsets in its action plans [28 (b)]

Disclosure dependent on materiality | Voluntary disclosure

ESRS E4 – Biodiversity and ecosystems (E4 – IV/XI)

Key disclosures concerning metrics and targets – 1/3

ESRS 2 MDR-T

Targets related to biodiversity and ecosystems [E4-4]

- Performance against each disclosed target, including an analysis of trends or significant changes in the performance [ESRS 2 MDR-T, 80 (j)]
- Information on how each target is monitored and reviewed and whether progress is in line with what had been initially planned [ESRS 2 MDR-T, 80 (j)]
- Whether the targets are informed by, and/or aligned with the Kunming-Montreal Global Biodiversity Framework, relevant aspects of the EU Biodiversity Strategy for 2030 and other biodiversity and ecosystem-related national policies and legislation [32 (b)]
- Whether biodiversity offsets are used in setting targets [32 (e)]
- To which layer in the mitigation hierarchy the target can be allocated [32 (f)]

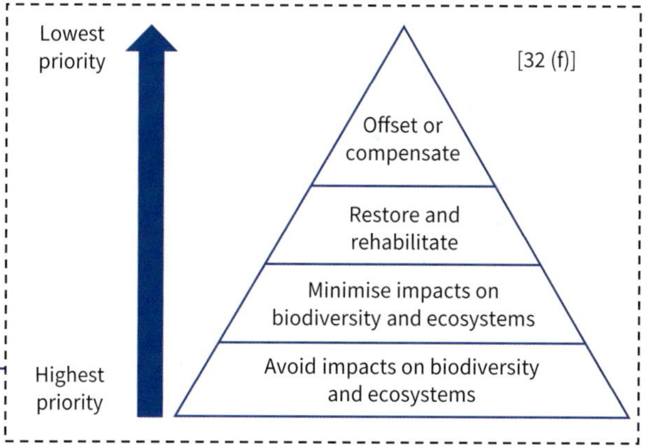

Disclosure dependent on materiality | Explanation/Examples

ESRS E4 – Biodiversity and ecosystems (E4 – V/XI)

Key disclosures concerning metrics and targets – 2/3

ESRS 2 MDR-M

Impact metrics related to biodiversity and ecosystems change [E4-5]

IX/XI

Metrics used to evaluate performance and effectiveness in relation to material IROs [ESRS 2 MDR-M, 75]

Whether the measurement of the metric is validated by an external body other than the assurance provider and, if so, which body [ESRS 2 MDR-M, 77 (b)]

Metrics concerning negatively affected sites located in or near protected and/or biodiversity-sensitive areas: [35]
- Number of relevant sites owned, leased or managed
- Area of relevant sites owned, leased or managed

Metrics concerning land-use:
- Land-use based on a life cycle assessment [36]
- Total use of land [AR 34 (a)]
- Total sealed area [AR 34 (b)]
- Total nature-oriented area on site [AR 34 (c)]
- Total nature-oriented area off site [AR 34 (d)]

Metrics concerning the impact drivers of land-use change, freshwater-use change and/or sea-use change that measure: [38]
- The conversion over time of land cover
- The functional connectivity
- Changes:
  - Over time in the management of the ecosystem
  - In the spatial configuration of the landscape
  - In ecosystem structural connectivity

Metrics concerning invasive alien species:
- Metrics used to manage pathways of introduction and spread of invasive alien species and risks posed by these [39]
- Number of invasive alien species [AR 32]
- Extent of surface covered by invasive alien species [AR 32]

Metrics concerning the state of species that measure: [40]
- The change in number of individuals of a species within a specific area
- The threat status of species and how activities/pressures may affect the threat status or changes in relevant habitat for a threatened species as proxy for the impact on extinction risk

Metrics concerning ecosystems that measure: [41]
- Area coverage of a particular ecosystem
- The quality of ecosystems relative to a pre-determined reference state
- Multiple species within an ecosystem, e.g. species richness and abundance indicators, aggregation of species' conservation status
- Structural components of condition such as habitat connectivity

Disclosure dependent on materiality | Voluntary disclosure

ESRS E4 – Biodiversity and ecosystems (E4 – VI/XI)

Key disclosures concerning metrics and targets – 3/3

ESRS 2 MDR-M

Anticipated financial effects [E4-6]

XI/XI

- Whether the measurement of the metric is validated by an external body other than the assurance provider and, if so, which body [ESRS 2 MDR-M, 77 (b)]
- Anticipated financial effects in monetary terms due to material risks before considering biodiversity and ecosystem-related actions [45 (a)]
- Anticipated financial effects in monetary terms due to material opportunities before considering biodiversity and ecosystem-related actions [45 (a)]
- Assessment of related products and services at risk over the short-, medium- and long-term, explaining how these are defined, how financial amounts are estimated and which critical assumptions are made [AR 39]

Disclosure dependent on materiality with phase-in

Voluntary disclosure

ESRS E4 – Biodiversity and ecosystems (E4 – VII/XI)

Disclosure Requirement ESRS E4-1 | Transition plan and consideration in strategy and business model– 1/2

Resilience assessment

Enable an understanding of the resilience of the undertaking's strategy and business model in relation to biodiversity and ecosystems and of the compatibility of the undertaking's strategy and business model with regard to relevant local, national and global public policy targets related to biodiversity and ecosystems. [12]

Input factors

Scope of the resilience analysis [13 (b)]

- Own operations
- Value chain
- Risks considered

Key assumptions made [13 (c)]

Time horizons used [13 (d)]

→

Assessment

Assessment of the resilience of the current business model and strategy to biodiversity and ecosystem-related risks [13 (a)]

- Physical risks
- Transition risks
- Systemic risks

→

Output factors

Results of the resilience analysis [13 (e)]

Involvement of stakeholders, including, where appropriate, holders of indigenous and local knowledge [13 (f)]

If information specified in this DR is disclosed as part of the information required under ESRS 2 SBM-3, the undertaking may refer to this. [14]

Disclosure dependent on materiality   Explanation/Examples

ESRS E4 – Biodiversity and ecosystems (E4 – VIII/XI)

Disclosure Requirement ESRS E4-1 | Transition plan and consideration in strategy and business model – 2/2

Transition plan

A transition plan is a specific type of action plan that is adopted in relation to a strategic decision and that addresses a public policy objective and/or an entity-specific action plan including targets and actions, key strategic decisions, a major change in business model and/or particularly important actions and allocated resources. [Annex II]

Disclosure of the transition plan to improve and, ultimately, achieve alignment of the business model and strategy with the vision of the Kunming-Montreal Global Biodiversity Framework and its relevant goals and targets, the EU Biodiversity Strategy for 2030 and with respecting Planetary Boundaries related to biosphere integrity and land-system change [15]

If a transition plan is disclosed, explain e.g.: [AR 1]

- How the undertaking will adjust its strategy and business model to improve and, ultimately, achieve alignment with relevant local, national and global public policy goals and targets and, as appropriate, Planetary Boundaries related to biosphere integrity and land-system change
- How the undertaking is responding to material impacts in its value chain
- How the undertaking contributes to addressing impact drivers and possible mitigation actions
- Quantification and explanation of the investments and funding supporting the implementation of the transition plan, with a reference to the key performance indicators of taxonomy-aligned CapEx, and where relevant the CapEx plans
- The use of biodiversity offsets
- How the undertaking measures progress, namely indicating the metrics and methodologies it uses for that purpose

If disclosing a transition plan, the undertaking may, for example, refer to the targets from the EU Biodiversity Strategy for 2030 [AR 2] or the SDGs [AR 3].

Voluntary disclosure | Explanation/Examples

ESRS E4 – Biodiversity and ecosystems (E4 – IX/XI)

Disclosure Requirement E4-5 | Impact metrics related to biodiversity and ecosystems change– 1/2

Overall disclosures

Summarised information on metrics and the methodologies and assumptions used as per ESRS 2 MDR-M [ESRS 2.73 ff.]

How the selected metrics correspond to the criteria of verifiability and being technically and scientifically robust considering the appropriate time scales geographies [AR 28]

Sites owned, leased or managed in or near protected and/or biodiversity-sensitive areas that are negatively affected by the undertaking [35]

- Number of sites
- Area in hectares

If the undertaking has identified material impacts with regards to land-use change, or impacts on the extent and condition of ecosystems, it may also disclose their land-use based on a life cycle assessment [36] with reference to specific indicators developed by the Joint Research Center [AR 31] and disclose in unit of area the total use of land, sealed area, nature-oriented area on site and nature-oriented area off site. [AR 34]

When preparing information required under E4-5, the undertaking shall consider and may describe: [AR 27]

- Methodologies and metrics used including contextual information
- The scope of the metrics and methodologies
- The biodiversity components of the metrics
- The geographies covered by the metrics including reasons for omitted geographies
- How the metrics integrate ecological thresholds and allocations
- Frequency of monitoring, key metrics being monitored, baseline value, base year and reference period
- On which kind of data the metrics rely (e.g. primary, secondary)
- Indication of which action is measured through the metric and how it relates to the achievement of targets
- Whether metrics are mandatory (required by legislation) or voluntary, the relevant legislation or voluntary standard or procedure
- Whether metrics are informed by or correspond to relevant authoritative national, EU-level or intergovernmental guidelines, policies or legislation

Disclosure dependent on materiality | Voluntary disclosure

ESRS E4 – Biodiversity and ecosystems (E4 – X/XI)

Disclosure Requirement E4-5 | Impact metrics related to biodiversity and ecosystems change – 2/2

Disclosure of specific metrics

Land-use, freshwater-use and/or sea-use change: relevant metrics that measure: [38]

- Conversion over time (e.g. 1 or 5 years) of land cover
- Changes over time (e.g. 1 or 5 years) in the management of ecosystems
- Changes in the spatial configuration of the landscape
- Changes in ecosystem structural connectivity
- The functional connectivity

State of species [40]

- Reference to the relevant DRs in ESRS E1, E2, E3 and E5 concerning relevant metrics
- Metrics measuring change in the number of individuals of a species within a specific area
- Metrics on species at extinction risk measuring the threat status and how it is affected or metrics measuring the change in habitat
- Consider population size, range within specific ecosystems as well as extinction risk. These aspects provide insight on the health of a single species' population and its relative resilience to human induced and naturally occurring change.

Invasive alien species

- Metrics used to manage pathways of introduction and spread of invasive alien species and risks posed by these [39]
- Pathways of invasive alien species [AR 32]
- Number of invasive alien species and extent of surface covered by these [AR 32]

Ecosystems [41]

Ecosystem extent

- Metrics that measure the area coverage of a particular ecosystem without necessarily considering the quality of the area being assessed, e.g. habitat cover

Ecosystem condition

- Metrics that measure the quality of ecosystems relative to pre-defined reference state
- Metrics that measure multiple species, e.g. richness and abundance indicators, aggregation of species' conservation status
- Metrics that reflect structural components of ecosystem condition, e.g. habitat connectivity

Note:
For datapoints specified in paragraphs 38 to 41 including the respective AR, the undertaking shall consider **its own operations**. [37] Resulting from ESRS 2 MDR-M, the undertaking shall disclose any metrics that it uses to evaluate performance and effectiveness in relation to a material IRO. The voluntary metrics included in ESRS E4-5 can be seen as examples for these metrics to be reported.

Voluntary disclosure

Explanation/Examples

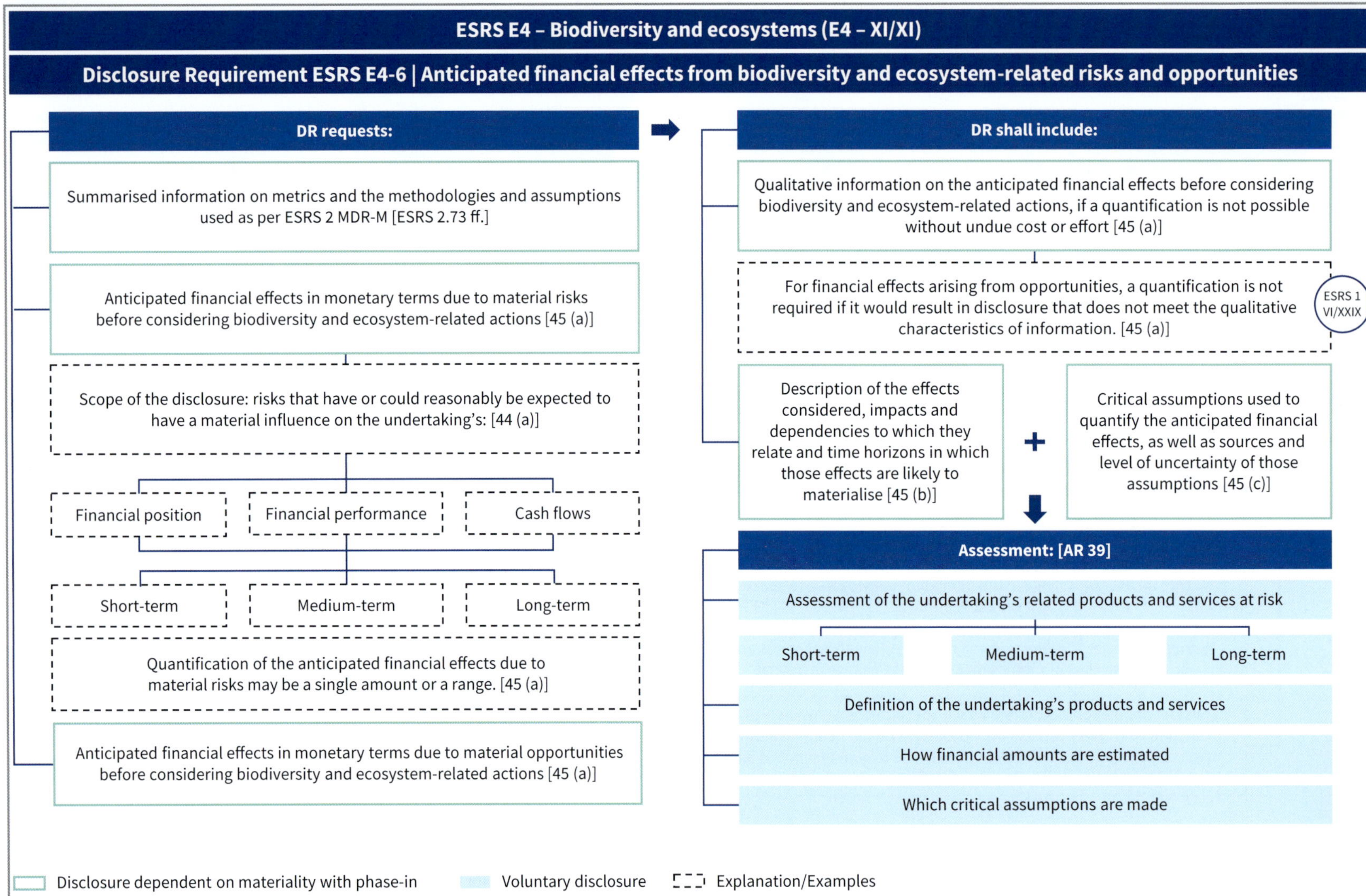
ESRS E4 – Biodiversity and ecosystems (E4 – XI/XI)
Disclosure Requirement ESRS E4-6 | Anticipated financial effects from biodiversity and ecosystem-related risks and opportunities
DR requests:
Summarised information on metrics and the methodologies and assumptions used as per ESRS 2 MDR-M [ESRS 2.73 ff.]
Anticipated financial effects in monetary terms due to material risks before considering biodiversity and ecosystem-related actions [45 (a)]
Scope of the disclosure: risks that have or could reasonably be expected to have a material influence on the undertaking's: [44 (a)]
Financial position
Financial performance
Cash flows
Short-term
Medium-term
Long-term
Quantification of the anticipated financial effects due to material risks may be a single amount or a range. [45 (a)]
Anticipated financial effects in monetary terms due to material opportunities before considering biodiversity and ecosystem-related actions [45 (a)]
DR shall include:
Qualitative information on the anticipated financial effects before considering biodiversity and ecosystem-related actions, if a quantification is not possible without undue cost or effort [45 (a)]
For financial effects arising from opportunities, a quantification is not required if it would result in disclosure that does not meet the qualitative characteristics of information. [45 (a)]
ESRS 1 VI/XXIX
Description of the effects considered, impacts and dependencies to which they relate and time horizons in which those effects are likely to materialise [45 (b)]
+
Critical assumptions used to quantify the anticipated financial effects, as well as sources and level of uncertainty of those assumptions [45 (c)]
Assessment: [AR 39]
Assessment of the undertaking's related products and services at risk
Short-term
Medium-term
Long-term
Definition of the undertaking's products and services
How financial amounts are estimated
Which critical assumptions are made
Disclosure dependent on materiality with phase-in
Voluntary disclosure
Explanation/Examples

ESRS E5 – Resource use and circular economy

Objective:

Enabling understanding of:

- The material positive and negative actual and potential impacts of the undertaking on resource use and circular economy, including resource efficiency, avoiding the depletion of resources and the sustainable sourcing and use of renewable resources.
- The plans and capacity of the undertaking to adapt its strategy and business model, in line with circular economy principles, including but not limited to minimising waste, maintaining the value of products, materials and other resources at their highest value and enhancing their efficient use in production and consumption.
- Any actions taken and the result of such actions to prevent or mitigate actual or potential negative impacts, including its measures to help decoupling its economic growth from the use of materials, and to address risks and opportunities.
- The nature, type and extent of the undertaking's material risks and opportunities arising from its impacts and dependencies related to resource use and circular economy and how it manages them.
- The financial effects on the undertaking over the short-, medium- and long-term of material risks and opportunities arising from its impacts and dependencies on resource use and circular economy.

Sustainability matters: Resource inflows including resource use, resource outflows related to products and services, waste

Interaction with other ESRS:

- **Conjunctions**: ESRS 1 (General requirements), ESRS 2 (General disclosures)
- **Intersectional topics covered by other standards**:
  ESRS E1 (GHG emissions and energy resources (energy consumption))
  ESRS E2 (Emissions to water, air and soil as well as substances of concern)
  ESRS E3 (Water resources (water consumption) and marine resources)
  ESRS E4 (Ecosystems, species and raw materials)
  ESRS S3 (Material negative impacts on affected communities from resource use and circular economy, in particular related to waste)

Interaction with other frameworks:

- Circular Economy Action Plan
- European Waste Catalogue
- Recommendation on methods for life cycle environmental performance
- Science-Based Targets Initiative for Nature
- Waste Framework Directive

Overall evaluation:

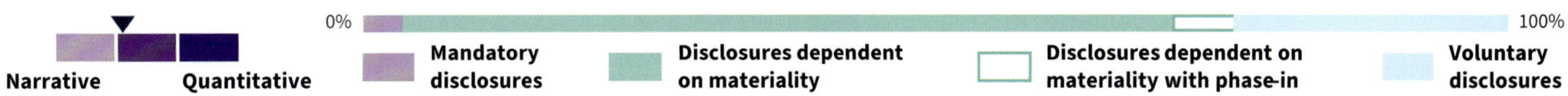

The overall evaluation contains a subjective assessment of the corresponding standard as narrative or quantitative and a count of all disclosures included in the corresponding standard. The number of disclosures was determined using professional judgement and depends on various factors. Due to the subjectivity involved, other counts may deviate.

ESRS E5 – Resource use and circular economy (E5 – I/VIII)

Structure and content

General disclosures

ESRS 2 IRO-1

ESRS 2 XVI/XXII

Description of the **processes** to **identify** and **assess** material resource use and circular economy-related **IROs**

Information on whether the undertaking has screened its assets and activities in order to identify IROs, including methodologies, assumptions and tools used and whether and how consultations were conducted, in particular, with affected communities. The LEAP approach may be considered.

Impact, risk and opportunity management

E5-1

ESRS 2 MDR-P

Policies related to resource use and circular economy

Information on the extent to which policies address the identification, assessment, management and/or remediation of material IROs related to resource use and circular economy.

E5-2

ESRS 2 MDR-A

Actions and **resources** related to resource use and circular economy

Information on key actions taken and planned and the resources allocated to their implementation to achieve resource use and circular economy-related policy objectives and targets.

Metrics and targets

E5-3

ESRS 2 MDR-T | V/VIII

Targets related to resource use and circular economy

Information on targets set to support resource use and circular economy-related policies and address material resource use and circular economy-related IROs.

E5-4

ESRS 2 MDR-M | VI/VIII

Resource inflows

Information on resource use in own operations and the upstream value chain. The material inflows include products (including packaging) and materials (critical raw materials and rare earths), water and property, plant and equipment used.

E5-5

ESRS 2 MDR-M | VII/VIII

Resource outflows

Information on the contribution of the undertaking to circular economy by designing products and materials in line with circular economy principles and by increasing or maximising the extent to which products, materials and waste processing are recirculated in practice after first use; as well as information on the waste reduction and waste management strategy and the knowledge on how waste is managed in own activities.

E5-6

ESRS 2 MDR-M | VIII/VIII

Anticipated financial effects from material resource use and circular economy-related risks and opportunities

Information on anticipated financial effects due to material risks arising from resource use and circular economy-related impacts and dependencies.

Information on anticipated financial effects due to material opportunities related to resource use and circular economy.

ESRS E5 – Resource use and circular economy (E5 – II/VIII)

Key disclosures concerning IRO management

ESRS 2 MDR-A

Actions and resources related to resource use and circular economy [E5-2]

- List of key actions taken and planned, their expected outcomes and their contribution to the achievement of policy objectives and targets [ESRS 2 MDR-A, 68 (a)]
- Key actions taken (along with results) to provide remedy for those harmed by actual material impacts [ESRS 2 MDR-A, 68 (d)]
- Whether the ability to implement each key action depends on specific preconditions [ESRS 2 MDR-A, 69 (a)]
- Amount of current and future financial resources for the implementation of a key action or action plan [ESRS 2 MDR-A, 69 (b), (c)]

Disclosure dependent on materiality

ESRS E5 – Resource use and circular economy (E5 – III/VIII)

Key disclosures concerning metrics and targets – 1/2

Targets related to resource use and circular economy [E5-3]

ESRS 2 MDR-T — V/VIII

- Performance against each disclosed target, including an analysis of trends or significant changes in the performance [ESRS 2 MDR-T, 80 (j)]
- Information on how each target is monitored and reviewed and whether progress is in line with what had been initially planned [ESRS 2 MDR-T, 80 (j)]
- To which layer of the waste hierarchy a target relates [25]

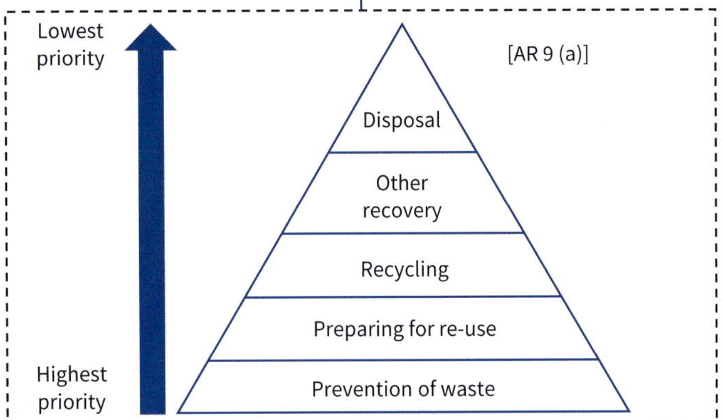

Resource inflows [E5-4]

ESRS 2 MDR-M — VI/VIII

- Whether the measurement of the metric is validated by an external body other than the assurance provider and, if so, which body [ESRS 2 MDR-M, 77 (b)]
- Weight of products and technical and biological materials used to manufacture products and services [31 (a)]
- Share of biological materials (and biofuels for non-energy purposes) used to manufacture products and services (including packaging) that are sustainably sourced [31 (b)]
- Weight and share of secondary reused or recycled components, secondary intermediary products and secondary materials used to manufacture products and services (including packaging) [31 (c)]

Disclosure dependent on materiality | Explanation/Examples

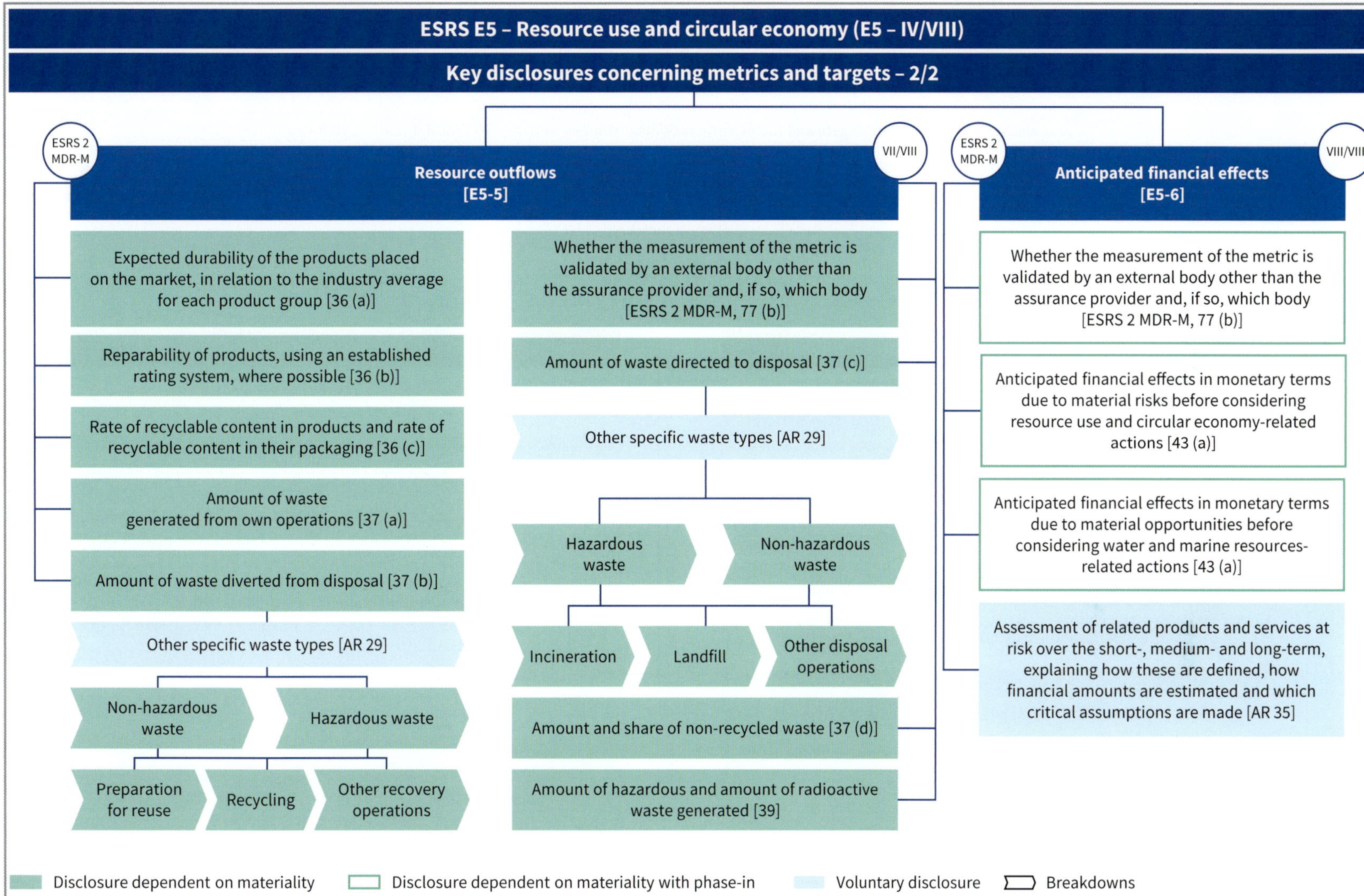
ESRS E5 – Resource use and circular economy (E5 – IV/VIII)
Key disclosures concerning metrics and targets – 2/2
ESRS 2 MDR-M
Resource outflows [E5-5]
VII/VIII
Expected durability of the products placed on the market, in relation to the industry average for each product group [36 (a)]
Reparability of products, using an established rating system, where possible [36 (b)]
Rate of recyclable content in products and rate of recyclable content in their packaging [36 (c)]
Amount of waste generated from own operations [37 (a)]
Amount of waste diverted from disposal [37 (b)]
Other specific waste types [AR 29]
Non-hazardous waste
Hazardous waste
Preparation for reuse
Recycling
Other recovery operations
Whether the measurement of the metric is validated by an external body other than the assurance provider and, if so, which body [ESRS 2 MDR-M, 77 (b)]
Amount of waste directed to disposal [37 (c)]
Other specific waste types [AR 29]
Hazardous waste
Non-hazardous waste
Incineration
Landfill
Other disposal operations
Amount and share of non-recycled waste [37 (d)]
Amount of hazardous and amount of radioactive waste generated [39]
ESRS 2 MDR-M
Anticipated financial effects [E5-6]
VIII/VIII
Whether the measurement of the metric is validated by an external body other than the assurance provider and, if so, which body [ESRS 2 MDR-M, 77 (b)]
Anticipated financial effects in monetary terms due to material risks before considering resource use and circular economy-related actions [43 (a)]
Anticipated financial effects in monetary terms due to material opportunities before considering water and marine resources-related actions [43 (a)]
Assessment of related products and services at risk over the short-, medium- and long-term, explaining how these are defined, how financial amounts are estimated and which critical assumptions are made [AR 35]
Disclosure dependent on materiality
Disclosure dependent on materiality with phase-in
Voluntary disclosure
Breakdowns

ESRS E5 – Resource use and circular economy (E5 – V/VIII)

Disclosure Requirement ESRS E5-3 | Targets related to resource use and circular economy

Summarised information on targets and the performance against those as per ESRS 2 MDR-T [ESRS 2.78 ff.]

To which layer of the waste hierarchy the target relates to [25] and whether targets set are mandatory (required by legislation) or voluntary [27]

Details on ecological thresholds [26]

Whether the target addresses shortcomings related to the Substantial Contribution criteria for Circular Economy and whether the target addresses shortcomings related to the Do No Significant Harm criteria [AR 20]

Whether and how targets relate to: [24]

- Resource inflows
- Resource outflows
- Waste

Including products and materials and, more specifically, to:

- The increase of circular product design (including e.g. design for durability, dismantling, reparability, recyclability) [24 (a)]
- The increase of the circular material use rate [24 (b)]
- The minimisation of primary raw material [24 (c)]
- Sustainable sourcing and use (in line with the cascading principle) of renewable resources [24 (d)]
- The waste management, including preparation for proper treatment [24 (e)]
- Other matters related to resource use or circular economy [24 (f)]

Further considerations:

- How targets may impact biodiversity loss. [AR 17]
- Targets in absolute value shall be prioritised. [AR 15]
- The production phase, the use phase and the end of functional life of products and materials shall be considered. [AR 16]
- Targets may cover own operations and/or value chain. [AR 19]

Explanation of the definition of sustainable sourcing the undertaking has adopted to disclose other targets and how the definition relates to the objective of the DR [AR 18]

Disclosure dependent on materiality | Voluntary disclosure | Explanation/Examples

ESRS E5 – Resource use and circular economy (E5 – VI/VIII)

Disclosure Requirement ESRS E5-4 | Resource inflows

Overall

Summarised information on metrics and the methodologies and assumptions used as per ESRS 2 MDR-M [ESRS 2.73 ff.]

Information on the calculation of the data: methodologies, sources and key assumptions [32]

Own operations and the upstream value chain

Description of resource inflows: [30]

- Products (including packaging)
- Materials (specifying critical raw materials and rare earths)
- Water
- Property, plant and equipment

used in own operations and along the upstream value chain

Own operations

Weight of products and technical and biological materials used (t or kg) [31 (a)]

Share of sustainably sourced biological materials (and biofuels for non-energy purposes) used to manufacture products and services (including packaging) and information on the certification scheme and application of the cascading principle [31 (b)]

Weight and share of secondary reused or recycled components, secondary intermediary products and secondary materials used to manufacture products and services (including packaging) (t or kg) [31 (c)]

Reported usage data are to reflect material in its original state without further data manipulation such as "dry weight". [AR 24]

Calculation of the share: [AR 23]

- Biological materials [31 (b)]
- Secondary reused or recycled components, etc. [31 (c)]

divided by # overall total weight of materials used, × 100

Disclosure dependent on materiality

Explanation/Examples

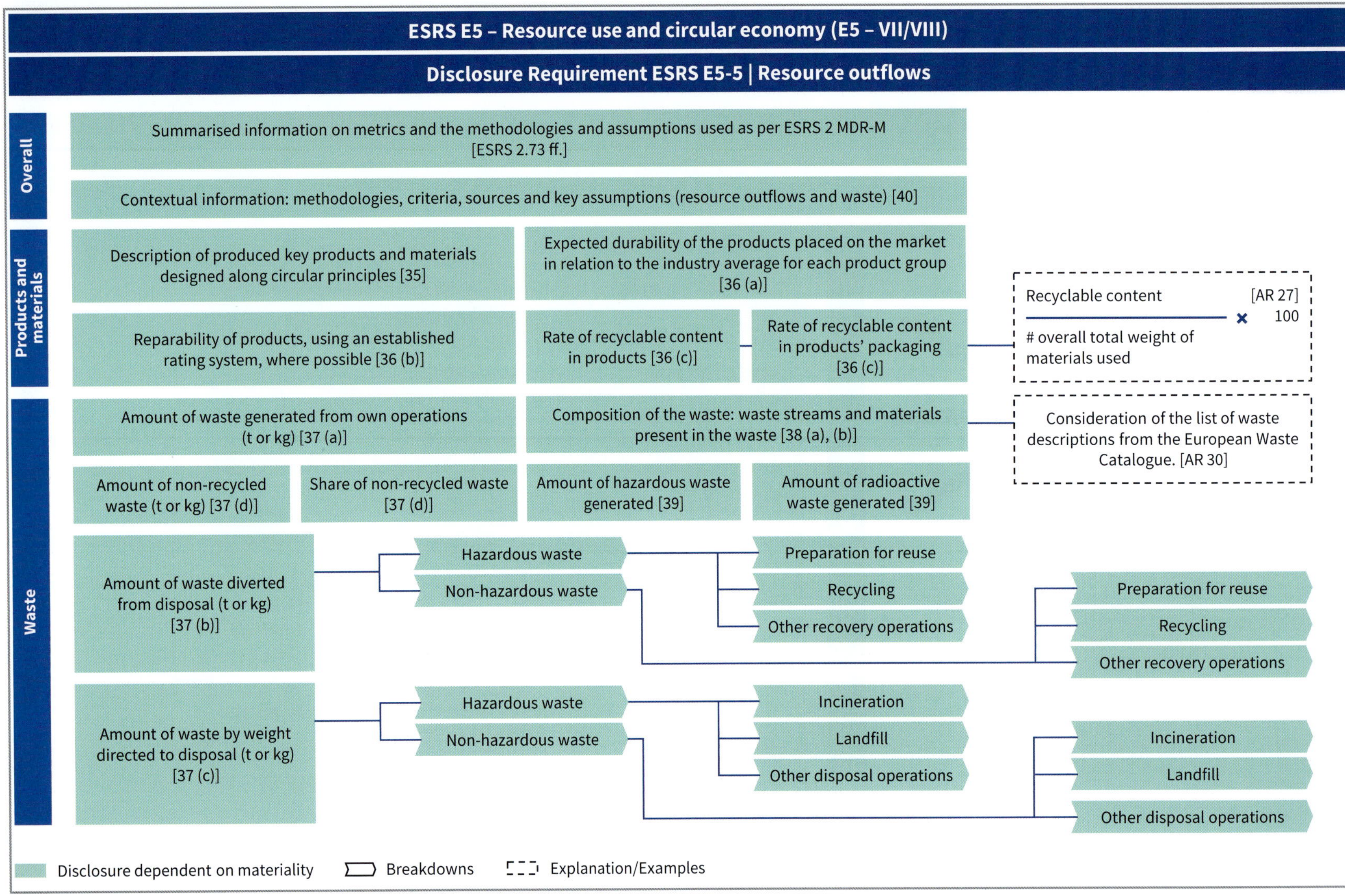
ESRS E5 – Resource use and circular economy (E5 – VII/VIII)
Disclosure Requirement ESRS E5-5 | Resource outflows
Overall
Summarised information on metrics and the methodologies and assumptions used as per ESRS 2 MDR-M [ESRS 2.73 ff.]
Contextual information: methodologies, criteria, sources and key assumptions (resource outflows and waste) [40]
Products and materials
Description of produced key products and materials designed along circular principles [35]
Expected durability of the products placed on the market in relation to the industry average for each product group [36 (a)]
Reparability of products, using an established rating system, where possible [36 (b)]
Rate of recyclable content in products [36 (c)]
Rate of recyclable content in products' packaging [36 (c)]
Recyclable content
overall total weight of materials used
× 100
[AR 27]
Waste
Amount of waste generated from own operations (t or kg) [37 (a)]
Composition of the waste: waste streams and materials present in the waste [38 (a), (b)]
Consideration of the list of waste descriptions from the European Waste Catalogue. [AR 30]
Amount of non-recycled waste (t or kg) [37 (d)]
Share of non-recycled waste [37 (d)]
Amount of hazardous waste generated [39]
Amount of radioactive waste generated [39]
Amount of waste diverted from disposal (t or kg) [37 (b)]
Hazardous waste
Non-hazardous waste
Preparation for reuse
Recycling
Other recovery operations
Preparation for reuse
Recycling
Other recovery operations
Amount of waste by weight directed to disposal (t or kg) [37 (c)]
Hazardous waste
Non-hazardous waste
Incineration
Landfill
Other disposal operations
Incineration
Landfill
Other disposal operations
Disclosure dependent on materiality
Breakdowns
Explanation/Examples

ESRS E5 – Resource use and circular economy (E5 – VIII/VIII)

Disclosure Requirement ESRS E5-6 | Anticipated financial effects

DR requests:

- Summarised information on metrics and the methodologies and assumptions used as per ESRS 2 MDR-M [ESRS 2.73 ff.]
- Anticipated financial effects in monetary terms due to material risks before considering resource use and circular economy-related actions [43 (a)]
  - Scope of the disclosure: risks that have or could reasonably be expected to have a material influence on the undertaking's: [42 (a)]
    - Financial position
    - Financial performance
    - Cash flows
      - Short-term
      - Medium-term
      - Long-term
  - Quantification of the anticipated financial effects due to material risks may be a single amount or a range. [AR 36]
- Anticipated financial effects in monetary terms due to material opportunities before considering resource use and circular economy-related actions [42 (b)]

DR shall include:

- Qualitative information on the anticipated financial effects before considering resource use and circular economy-related actions, if a quantification is not possible without undue cost or effort [43 (a)]
  - For financial effects arising from opportunities, a quantification is not required if it would result in disclosure that does not meet the qualitative characteristics of information. [43 (a)] (ESRS 1 VI/XXIX)
- Description of the effects considered, impacts and dependencies to which they relate and time horizons in which those effects are likely to materialise [43 (b)] + Critical assumptions used to quantify the anticipated financial effects, as well as sources and level of uncertainty of those assumptions [43 (c)]
- Illustration and description of how the undertaking intends to strengthen value retention with regards to opportunities [AR 34]

Assessment: [AR 35]

- Assessment of the undertaking's related products and services at risk
  - Short-term
  - Medium-term
  - Long-term
- Definition of the undertaking's products and services
- How financial amounts are estimated
- Which critical assumptions are made

Legend: Disclosure dependent on materiality with phase-in | Voluntary disclosure | Explanation/Examples

6 Social

ESRS S1 – S4 (I/III)

Overview of sustainability matters of ESRS S-Standards – 1/2

Overview of ESRS IX/XIII

Topics: Own workforce S1

S1-I/XIII

Sub-topics / Sub-sub-topics:

Working conditions

- Secure employment
- Working time
- Adequate wages
- Social dialogue
- Freedom of association, works councils and rights of workers
- Collective bargaining
- Work-life balance
- Health and safety

Equal treatment and opportunities for all

- Gender equality and equal pay for work of equal value
- Training and skills development
- Employment and inclusion of persons with disabilities
- Measures against violence and harassment in the workplace
- Diversity

Other work-related rights

- Child labour
- Forced labour
- Adequate housing
- Privacy

Topics: Workers in the value chain S2

S2-I/IX

Sub-topics / Sub-sub-topics:

Working conditions

- Secure employment
- Working time
- Adequate wages
- Social dialogue
- Freedom of association and work councils
- Collective bargaining
- Work-life balance
- Health and safety

Equal treatment and opportunities for all

- Gender equality and equal pay for work of equal value
- Training and skills development
- The employment and inclusion of persons with disabilities
- Measures against violence and harassment in the workplace
- Diversity

Other work-related rights

- Child labour
- Forced labour
- Adequate housing
- Water and sanitation
- Privacy

ESRS S1 – S4 (II/III)

Overview of sustainability matters of ESRS S-Standards – 2/2

Overview of ESRS X/XIII

Topics | **Sub-topics** | **Sub-sub-topics**

Affected communities S3

S3-I/IX

Communities' economic, social and cultural rights

- Adequate housing
- Adequate food
- Water and sanitation
- Land-related impacts
- Security-related impacts

Communities' civil and political rights

- Freedom of expression
- Freedom of assembly
- Impacts on human rights defenders

Rights of indigenous peoples

- Free, prior and informed consent
- Self-determination
- Cultural rights

Consumers and end-users S4

S4-I/IX

Information-related impacts for consumers and/or end-users

- Privacy
- Freedom of expression
- Access to (quality) information

Personal safety of consumers and/or end-users

- Health and safety
- Security of a person
- Protection of children

Social inclusion of consumers and/or end-users

- Non-discrimination
- Access to products and services
- Responsible marketing practices

ESRS S1 – S4 (III/III)

Cross-reference between ESRS 2 and ESRS S-Standards

ESRS Standard		General disclosures						IRO management		Metrics and targets	
		Topical disclosures to be reported alongside ESRS 2 disclosures						MDRs to be applied alongside topical ESRS			
		Governance		Strategy		Processes	Transition plan	Policies	Actions and resources	Metrics	Targets
		GOV-1	GOV-3	SBM-2	SBM-3	IRO-1		MDR-P	MDR-A	MDR-M	MDR-T
ESRS Standard	**S1** Own workforce			S1-I/XIII	S1-II/XIII			S1-III/XIII	S1-III/XIII	S1-IV/XIII S1-XI/XIII	S1-IV/XIII
	S2 Workers in the value chain			S2-I/IX	S2-II/IX			S2-III/IX S2-VI/IX	S2-IV/IX S2-IX/IX	S2-V/IX	S2-V/IX
	S3 Affected communities			S3-I/IX	S3-II/IX			S3-III/IX S3-VI/IX	S3-IV/IX S3-IX/IX	S3-V/IX	S3-V/IX
	S4 Consumers and end-users			S4-I/IX	S4-II/IX			S4-III/IX S4-VI/IX	S4-IV/IX S4-IX/IX	S4-V/IX	S4-V/IX

ESRS S1 – Own workforce

Objective:

Enabling understanding of:

- The material positive and negative actual and potential impacts of the undertaking on its own workforce.
- Any actions taken and the result of such actions to prevent, mitigate or remediate actual or potential negative impacts and to address risks and opportunities.
- The nature, type and extent of the undertaking's material risks and opportunities arising from its impacts and dependencies on its own workforce and how it manages them.
- The financial effects on the undertaking over the short-, medium- and long-term of material risks and opportunities arising from its impacts and dependencies on its own workforce.
- The extent to which the undertaking aligns or complies with international and European human rights instruments and conventions.

Sustainability matters:

Working conditions (secure employment, working time, adequate wages, social dialogue, freedom of association, the existence of works councils and the information, consultation and participation rights of workers, collective bargaining including the rate of workers covered by collective agreements, work-life balance, health and safety), equal treatment and opportunities for all (gender equality and equal pay for work of equal value, training and skills development, employment and inclusion of persons with disabilities, measures against violence and harassment in the workplace, diversity), other work-related rights (child labour, forced labour, adequate housing, privacy)

Interaction with other ESRS:

- **Conjunctions**: ESRS 1 (General requirements), ESRS 2 (General disclosures), ESRS S2 (Workers in the value chain), ESRS S3 (Affected communities), ESRS S4 (Consumers and end-users)
- **Intersectional topics covered by other standards**: n/a

Interaction with other frameworks:

- Directive on minimum wages
- European Convention of Human Rights
- European Pillar of Social Rights
- European Social Charter (revised)
- ILO Declaration on Fundamental Principles and Rights at Work
- ILO Fundamental Conventions
- ILO List of Occupational Diseases
- International Bill of Human Rights
- OECD Guidelines for Multinational Enterprises
- The Charter of Fundamental Rights of the European Union
- UN Convention on Rights of Persons with Disabilities
- UNGPs on Business and Human Rights

Applicability:

Undertakings or groups not exceeding the average number of 750 employees may omit the information specified in the DRs of ESRS S1 for the first year of preparation of their sustainability statement. In this case the undertaking discloses the requirements according to ESRS 2.17.

ESRS 2 X/XXII

Overall evaluation:

The overall evaluation contains a subjective assessment of the corresponding standard as narrative or quantitative and a count of all disclosures included in the corresponding standard. The number of disclosures was determined using professional judgement and depends on various factors. Due to the subjectivity involved, other counts may deviate.

ESRS S1 – Own workforce (S1 – I/XIII)

Structure and content

General disclosures

ESRS 2 SBM-2

Interests and views of stakeholders

Information on how the interests, views and rights of (actually and potentially) materially affected people in the undertaking's own workforce, including respect for their human rights, inform the undertaking's strategy and business model.

ESRS 2 SBM-3

Material IROs and their interaction with **strategy and business model**

Whether and how IROs originate from and their relationship to strategy and business model, whether all people in the undertaking's own workforce are included in the disclosure as well as disclosure of e.g. operations at significant risk of child, forced or compulsory labour.

Impact, risk and opportunity management

S1-1 (ESRS 2 MDR-P)

Policies related to own workforce

Information on the extent to which policies address the identification, assessment, management and/or remediation of material IROs related to the undertaking's own workforce.

S1-2

Processes for engaging with own workforce and workers' representatives about impacts

Information on the general processes for engaging with people in the undertaking's own workforce and their representatives about actual and potential material impacts concerning the undertaking's own workforce.

S1-3

Processes to remediate negative impacts and **channels** for own workforce to raise concerns

Information on the processes in place to provide for or cooperate in the remediation of negative impacts on people in the undertaking's own workforce, as well as channels available to raise concerns and have them addressed.

S1-4 (ESRS 2 MDR-A)

Actions and **resources** for managing material IROs and the effectiveness of those actions

Information on any actions and initiatives aimed at preventing, mitigating and remedying negative material impacts; achieving positive material impacts; addressing material risks and pursuing material opportunities related to the undertaking's own workforce, including the effectiveness of those actions.

Metrics and targets

S1-5 (ESRS 2 MDR-T)

Targets related to own workforce

Information on targets set to reduce negative impacts, advance positive impacts and/or manage material risks and opportunities related to the undertaking's own workforce.

S1-6 Characteristics of employees	**S1-7** Characteristics of non-employees	**S1-8** Collective bargaining coverage and social dialogue
S1-9 Diversity metrics	**S1-10** (XI/XIII) Adequate wages	**S1-11** Social protection
S1-12 Persons with disabilities	**S1-13** Training and skills develop-ment metrics	**S1-14** (XII/XIII) Health and safety metrics
S1-15 Work-life balance metrics	**S1-16** (XIII/XIII) Remune-ration metrics	**S1-17** Incidents, complaints and severe human rights impacts

ESRS 2 MDR-M

ESRS S1 – Own workforce (S1 – II/XIII)

Key disclosures concerning general disclosures

Material IROs and their interaction with strategy and business model [ESRS 2 SBM-3]

- Whether all people in the undertaking's own workforce who could be materially impacted through the undertaking's own operations and the value chain are included in the scope of disclosure under ESRS 2 [14]
- Whether material negative impacts are either: [14 (b)]
  - Widespread or systemic in the undertaking's operating contexts or
  - Related to individual incidents
- Any material risks and opportunities arising from impacts and dependencies on the undertaking's own workforce [14 (d)]
- Any material impacts on the undertaking's own workforce that may arise from transition plans for reducing negative impacts on the environment [14 (e)]
- Operations at significant risk of incidents of forced, compulsory or child labour either in terms of: [14 (f), (g)]
  - Type of operation or
  - Countries or geographic areas with operations considered at risk
- Which, if any, material risks and opportunities arising from impacts and dependencies relate to specific groups of people rather than all people in the undertaking's own workforce [16]

Disclosure dependent on materiality

ESRS S1 – Own workforce (S1 – III/XIII)

Key disclosures concerning IRO management

ESRS 2 MDR-P

Policies related to own workforce [S1-1]

- Whether and how policies are aligned with internationally recognised standards [21]
- Statement on the alignment with instruments such as the International Bill of Human Rights and the International Labour Organisation (ILO) Declaration on Fundamental Principles and Rights at Work [AR 12]

Processes to remediate negative impacts and channels for own workforce to raise concerns [S1-3]

- Relevant and reliable data about the effectiveness of the channels to raise concerns from the perspective of the people concerned [AR 31]

ESRS 2 MDR-A

Actions and resources for managing material IROs and the effectiveness of those actions [S1-4]

- List of key actions taken and planned, their expected outcomes and their contribution to the achievement of policy objectives and targets [ESRS 2 MDR-A, 68 (a)]
- Key actions taken (along with results) to provide remedy for those harmed by actual material impacts [ESRS 2 MDR-A, 68 (d)]
- Whether the ability to implement each key action depends on specific preconditions [ESRS 2 MDR-A, 69 (a)]
- Amount of current and future financial resources for the implementation of a key action or action plan [ESRS 2 MDR-A, 69 (b), (c)]
- Whether and how the undertaking ensures that own practices do not cause or contribute to material negative impacts on the own workforce, including, where relevant, practices in relation to: [41]
  - Procurement
  - Sales
  - Data use

Disclosure dependent on materiality | Voluntary disclosure

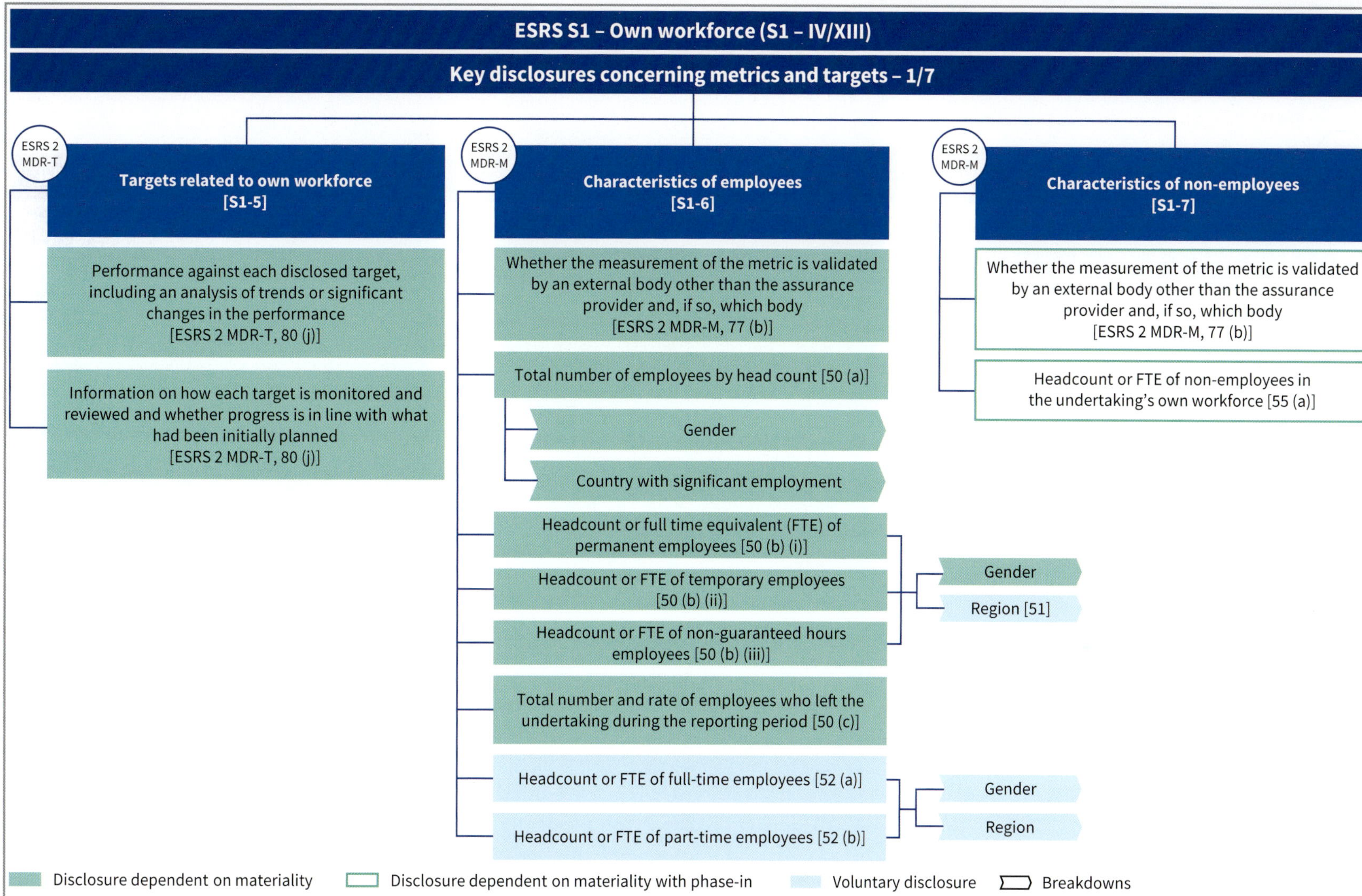
ESRS S1 – Own workforce (S1 – IV/XIII)
Key disclosures concerning metrics and targets – 1/7
ESRS 2 MDR-T
Targets related to own workforce [S1-5]
Performance against each disclosed target, including an analysis of trends or significant changes in the performance [ESRS 2 MDR-T, 80 (j)]
Information on how each target is monitored and reviewed and whether progress is in line with what had been initially planned [ESRS 2 MDR-T, 80 (j)]
ESRS 2 MDR-M
Characteristics of employees [S1-6]
Whether the measurement of the metric is validated by an external body other than the assurance provider and, if so, which body [ESRS 2 MDR-M, 77 (b)]
Total number of employees by head count [50 (a)]
Gender
Country with significant employment
Headcount or full time equivalent (FTE) of permanent employees [50 (b) (i)]
Headcount or FTE of temporary employees [50 (b) (ii)]
Headcount or FTE of non-guaranteed hours employees [50 (b) (iii)]
Gender
Region [51]
Total number and rate of employees who left the undertaking during the reporting period [50 (c)]
Headcount or FTE of full-time employees [52 (a)]
Headcount or FTE of part-time employees [52 (b)]
Gender
Region
ESRS 2 MDR-M
Characteristics of non-employees [S1-7]
Whether the measurement of the metric is validated by an external body other than the assurance provider and, if so, which body [ESRS 2 MDR-M, 77 (b)]
Headcount or FTE of non-employees in the undertaking's own workforce [55 (a)]
Disclosure dependent on materiality
Disclosure dependent on materiality with phase-in
Voluntary disclosure
Breakdowns

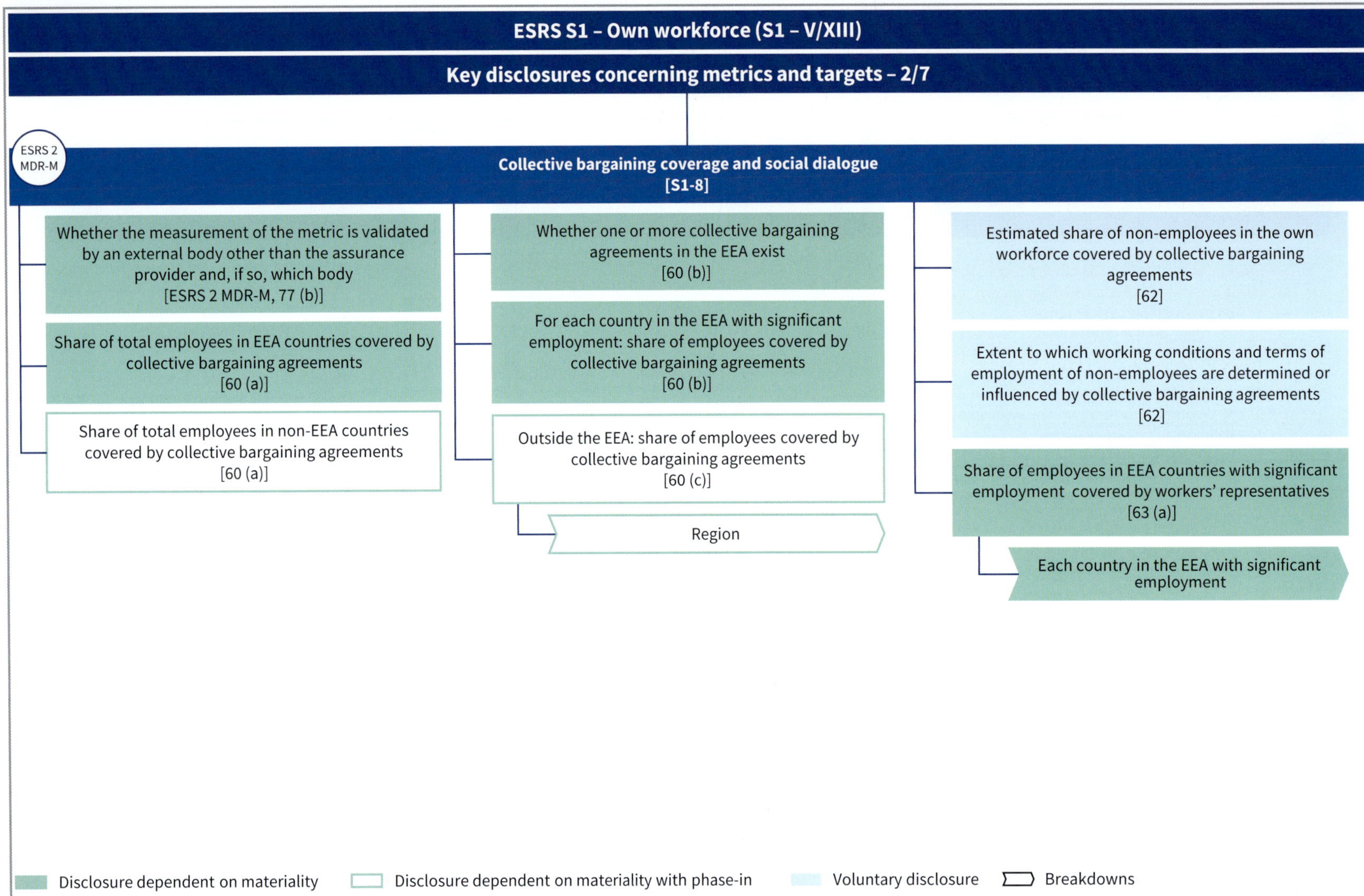
ESRS S1 – Own workforce (S1 – V/XIII)
Key disclosures concerning metrics and targets – 2/7
ESRS 2 MDR-M
Collective bargaining coverage and social dialogue [S1-8]
Whether the measurement of the metric is validated by an external body other than the assurance provider and, if so, which body [ESRS 2 MDR-M, 77 (b)]
Share of total employees in EEA countries covered by collective bargaining agreements [60 (a)]
Share of total employees in non-EEA countries covered by collective bargaining agreements [60 (a)]
Whether one or more collective bargaining agreements in the EEA exist [60 (b)]
For each country in the EEA with significant employment: share of employees covered by collective bargaining agreements [60 (b)]
Outside the EEA: share of employees covered by collective bargaining agreements [60 (c)]
Region
Estimated share of non-employees in the own workforce covered by collective bargaining agreements [62]
Extent to which working conditions and terms of employment of non-employees are determined or influenced by collective bargaining agreements [62]
Share of employees in EEA countries with significant employment covered by workers' representatives [63 (a)]
Each country in the EEA with significant employment
Disclosure dependent on materiality
Disclosure dependent on materiality with phase-in
Voluntary disclosure
Breakdowns

ESRS S1 – Own workforce (S1 – VI/XIII)

Key disclosures concerning metrics and targets – 3/7

Diversity metrics [S1-9] (ESRS 2 MDR-M)

- Whether the measurement of the metric is validated by an external body other than the assurance provider and, if so, which body [ESRS 2 MDR-M, 77 (b)]
- Gender distribution at top management level in number and share [66 (a)]
  - Gender
- Age distribution of employees (number and share) [66 (b)]
  - < 30 years old
  - 30-50 years old
  - > 50 years old

Adequate wages [S1-10] (ESRS 2 MDR-M; XI/XIII)

- Whether the measurement of the metric is validated by an external body other than the assurance provider and, if so, which body [ESRS 2 MDR-M, 77 (b)]
- Share of employees paid below applicable adequate wage benchmarks [70]
  - Each country concerned
- Share of non-employees in the own workforce paid below applicable adequate wage benchmarks [71]
  - Each country concerned

Disclosure dependent on materiality | Voluntary disclosure | Breakdowns

ESRS S1 – Own workforce (S1 – VII/XIII)

Key disclosures concerning metrics and targets – 4/7

Social protection [S1-11] (ESRS 2 MDR-M)

- Whether the measurement of the metric is validated by an external body other than the assurance provider and, if so, which body [ESRS 2 MDR-M, 77 (b)]
- Whether all employees are covered by social protection through public programs or through benefits offered by the undertaking against loss of income due to the following major life events: [74]
  - Sickness
  - Unemployment
  - Employment injury or acquired disability
  - Parental leave and
  - Retirement
- Whether all non-employees in own workforce are covered by social protection through public programs or through benefits offered by the undertaking against loss of income due to the following major life events: [76]
  - Sickness
  - Unemployment
  - Employment injury or acquired disability
  - Parental leave and
  - Retirement

Persons with disabilities [S1-12] (ESRS 2 MDR-M)

- Whether the measurement of the metric is validated by an external body other than the assurance provider and, if so, which body [ESRS 2 MDR-M, 77 (b)]
- Share of persons with disabilities amongst employees [79]
  - Gender [80]

Disclosure dependent on materiality with phase-in | Voluntary disclosure | Breakdowns

ESRS S1 – Own workforce (S1 – VIII/XIII)

Key disclosures concerning metrics and targets – 5/7

ESRS 2 MDR-M

Training and skills development metrics [S1-13]

- Whether the measurement of the metric is validated by an external body other than the assurance provider and, if so, which body [ESRS 2 MDR-M, 77 (b)]
- Share of employees that participated in regular performance and career development reviews [83 (a)]
- Number/proportion of performance reviews per employee [AR 77 (a)]
- Number of reviews concerning employees in proportion to the agreed number of reviews by the management [AR 77 (b)]
- Average number of training hours per employee [83 (b)]

Breakdowns:
- Gender
- Employee category [84]

- Share of non-employees in the own workforce that participated in regular performance and career development reviews [83 (a), 85]
- Number/proportion of performance reviews per non-employee in the own workforce [85, AR 77 (a)]
- Number of reviews concerning non-employees in the own workforce in proportion to the agreed number of reviews by the management [85, AR 77 (b)]
- Average number of training hours per non-employee in the own workforce [83 (b), 85]

Breakdowns:
- Gender
- Employee category [84, 85]

Disclosure dependent on materiality with phase-in | Voluntary disclosure | Breakdowns

ESRS S1 – Own workforce (S1 – IX/XIII)

Key disclosures concerning metrics and targets – 6/7

ESRS 2 MDR-M

Health and safety metrics [S1-14]

XII/XIII

- Whether the measurement of the metric is validated by an external body other than the assurance provider and, if so, which body [ESRS 2 MDR-M, 77 (b)]
- Share of people in the own workforce covered by the health and safety management (HSM) system [88 (a)]
- Number of fatalities due to work-related injuries and ill-health [88 (b)]
- Number and rate of recordable work-related accidents [88 (c)]
  - Employees
  - Non-employees
- Number of fatalities due to work-related injuries and ill-health among other workers working on the undertaking's sites (e.g. value chain workers) [88]
- Number of cases of recordable work-related ill-health with regard to employees [88 (d)]
- Number of days lost to work-related injuries and fatalities from work-related accidents, work-related ill-health and fatalities from ill-health with regard to employees [88 (e)]
- Number of cases of recordable work-related ill-health with regard to non-employees [89]
- Number of days lost to work-related injuries and fatalities from work-related accidents, work-related ill-health and fatalities from ill-health with regard to non-employees [89]
- Whether the HSM system, or certain parts of it, has been subject to an internal audit or external certification and the underlying standards for such audits/certifications [AR 81]
- Share of own workforce covered by an audited/certified HSM [90]
- Incidents which arise during travel, outside of the undertaking's responsibility [AR 84]

Disclosure dependent on materiality | Disclosure dependent on materiality with phase-in | Voluntary disclosure | Breakdowns

ESRS S1 – Own workforce (S1 – X/XIII)

Key disclosures concerning metrics and targets – 7/7

Work-life balance metrics [S1-15]
ESRS 2 MDR-T

- Whether the measurement of the metric is validated by an external body other than the assurance provider and, if so, which body [ESRS 2 MDR-M, 77 (b)]
- Share of employees entitled to take family-related leave [93 (a)]
- Share of entitled employees that took family-related leave [93 (b)]
  - Gender

Remuneration metrics [S1-16]
ESRS 2 MDR-M; XIII/XIII

- Whether the measurement of the metric is validated by an external body other than the assurance provider and, if so, which body [ESRS 2 MDR-M, 77 (b)]
- Gender pay gap [97 (a)]
  - Employee category [98]
  - Remuneration components [98]
  - Country or segment [98]
- Annual total remuneration ratio [97 (b)]
- Annual total remuneration ratio adjusted for purchasing power differences between countries [99]

Incidents, complaints and severe human rights impacts [S1-17]
ESRS 2 MDR-M

- Whether the measurement of the metric is validated by an external body other than the assurance provider and, if so, which body [ESRS 2 MDR-M, 77 (b)]
- Number of incidents of discrimination including harassment [103 (a)]
- Number of complaints filed through channels to raise concerns [103 (b)]
- Amount of fines, penalties and compensation for damages related to incidents under [103 (a), (b)] [103 (c)]
- Number of severe human rights issues and incidents and number of those that are cases of non-respect of relevant internationally recognised standards [104 (a)]
- Amount of fines, penalties and compensation for damages related to incidents under [104 (a)] [104 (b)]
- Number of severe human rights incidents where the undertaking played a role in securing remedy for those affected [AR 106]

Disclosure dependent on materiality | Disclosure dependent on materiality with phase-in | Voluntary disclosure | Breakdowns

ESRS S1 – Own workforce (S1 – XI/XIII)

Disclosure Requirement ESRS S1-10 | Adequate wages

Disclosures

Summarised information on metrics and the methodologies and assumptions used as per ESRS 2 MDR-M [ESRS 2.73 ff.]

Are all **employees** paid an adequate wage, in line with applicable benchmarks?

- **no** → Statement on not paying adequate wages to all employees [69]
  - + Share of employees paid below adequate wage [70]
    - Each country concerned
- **yes** → Statement on paying adequate wages to all employees [69]

Are all **non-employees in own workforce** paid an adequate wage, in line with applicable benchmarks?

- **no** → Statement on not paying adequate wages to all non-employees in own workforce [71]
  - + Share of non-employees paid below adequate wage [71]
    - Each country concerned
- **yes** → Statement on paying adequate wages to all non-employees in own workforce [71]

Additional information

How shall undertakings determine whether wages paid to employees and/or non-employee workers are adequate? [AR 72]

1. Calculation of the **lowest wage for the lowest pay category**
   - Excluding interns and apprentices
   - Based on basic wage plus any fixed additional payments that are guaranteed to all (non-) employees
   - Separately for each EEA country and each non-EEA region

2. Comparison of the lowest wage with the **adequate wage benchmark**

Definitions

Adequate wage:
Wage that provides for the satisfaction of the needs of the worker and his/her family in the light of national economic and social conditions. [Annex II]

Adequate wage benchmark:

≥

In the EEA:
Minimum wage set in accordance with the Directive on minimum wages; minimum wage determined by legislation or collective bargaining; minimum wage in a neighbouring country, commonly-referenced international norm (e.g. obtained from the European Labour Force Survey)
[AR 73 (a), AR 74]

Outside the EEA:
Wage level established in any existing international, national or sub-national legislation, official norms or collective agreements;
any national or sub-national minimum wage established by legislation or collective bargaining; benchmarks in line with the Sustainable Trade Initiative, Wage Indicator Foundation or Fair Wage Network [AR 73 (b)]

Disclosure dependent on materiality | Voluntary disclosure | Breakdowns | Explanation/Examples

ESRS S1 – Own workforce (S1 – XII/XIII)

Disclosure Requirement ESRS S1-14 | Health and safety metrics

Disclosures

Summarised information on metrics and the methodologies and assumptions used as per ESRS 2 MDR-M [ESRS 2.73 ff.]

Coverage of health and safety management

- Share of own workforce covered by an audited/certified HSM system [90]
- Share of own workforce covered by an HSM system [88 (a)]

Quality and performance of health and safety management

- Number of fatalities due to work-related injuries and ill-health in own workforce [88 (b)]
- Number and rate of recordable work-related accidents [88 (c)]
- Number of cases of recordable work-related ill-health with regard to employees [88 (d)]
- Number of days lost to work-related injuries and fatalities from work-related accidents, work-related ill-health and fatalities from ill-health with regard to employees [88 (e)]

Employees

Non-employee workers

- Number of fatalities as result of work-related injuries and work-related ill-health among other workers working on the undertaking's sites [88]
- Number of cases of recordable work-related ill-health with regard to non-employees [89]
- Number of days lost to work-related injuries and fatalities from work-related accidents, work-related ill-health and fatalities from ill-health with regard to non-employees [89]

Additional information

What incidents are considered "work-related"? [AR 83-88]
= accidents, injuries and ill-health arising from exposure to hazards at work

+ **Home office:** if related to performance of work rather than the general home environment
+ **Travel:** if worker was engaged in work activities in the interest of the employer at the time of the incident or if the undertaking was responsible for commute
+ **Mental illness**: if notified voluntarily by the person concerned, confirmed by licensed healthcare professional and diagnosed as related to work

Calculation of the **accident rate**: [AR 89]

$$\frac{\text{\# of accidents}}{\text{\# of total hours worked by own workforce}} \times 1{,}000{,}000$$

E.g. **acute, recurring and chronic health problems** caused or aggravated by work conditions or practices such as musculoskeletal disorders, skin and respiratory diseases, mental illnesses etc., at a minimum those cases outlined in the **ILO List of Occupational Diseases.** [AR 92]

Included in the number of **days lost**: [AR 95]

+ First full day and last day of absence
+ Calendar days on which affected individual is not scheduled for work (including public holidays and weekends)

Legend: Disclosure dependent on materiality | Disclosure dependent on materiality with phase-in | Voluntary disclosure | Breakdowns | Explanation/Examples

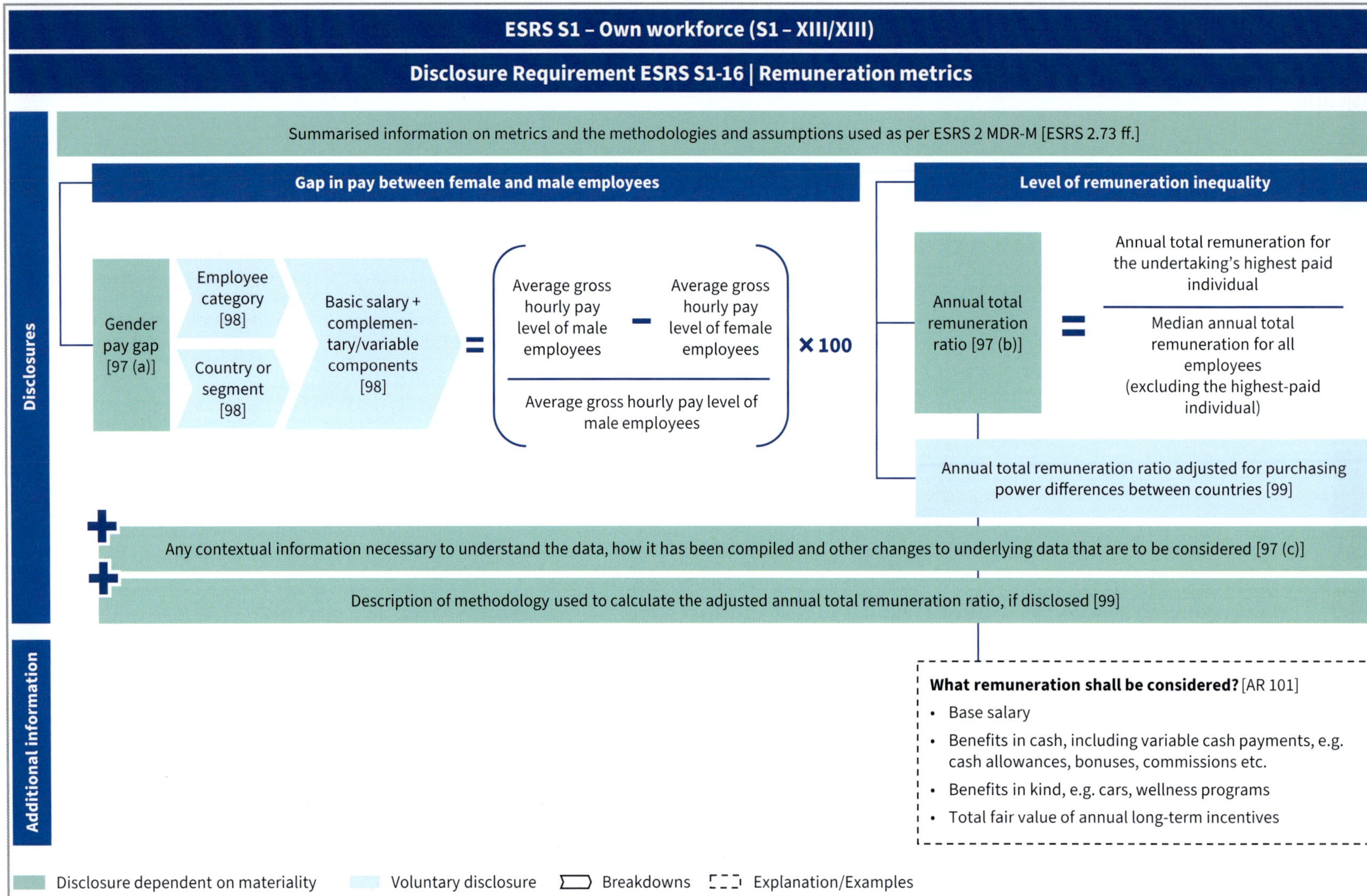
ESRS S1 – Own workforce (S1 – XIII/XIII)
Disclosure Requirement ESRS S1-16 | Remuneration metrics
Disclosures
Summarised information on metrics and the methodologies and assumptions used as per ESRS 2 MDR-M [ESRS 2.73 ff.]
Gap in pay between female and male employees
Gender pay gap [97 (a)]
Employee category [98]
Country or segment [98]
Basic salary + complementary/variable components [98]
=
Average gross hourly pay level of male employees
−
Average gross hourly pay level of female employees
Average gross hourly pay level of male employees
× 100
Level of remuneration inequality
Annual total remuneration ratio [97 (b)]
=
Annual total remuneration for the undertaking's highest paid individual
Median annual total remuneration for all employees (excluding the highest-paid individual)
Annual total remuneration ratio adjusted for purchasing power differences between countries [99]
Any contextual information necessary to understand the data, how it has been compiled and other changes to underlying data that are to be considered [97 (c)]
Description of methodology used to calculate the adjusted annual total remuneration ratio, if disclosed [99]
Additional information
What remuneration shall be considered? [AR 101]
• Base salary
• Benefits in cash, including variable cash payments, e.g. cash allowances, bonuses, commissions etc.
• Benefits in kind, e.g. cars, wellness programs
• Total fair value of annual long-term incentives
Disclosure dependent on materiality
Voluntary disclosure
Breakdowns
Explanation/Examples

ESRS S2 – Workers in the value chain

Objective:	Enabling understanding of: • The material positive and negative actual and potential impacts of the undertaking on workers in its value chain. • Any actions taken and the result of such actions to prevent, mitigate or remediate actual or potential negative impacts and to address risks and opportunities. • The nature, type and extent of the undertaking's material risks and opportunities arising from its impacts and dependencies on workers in the value chain and how it manages them. • The financial effects on the undertaking over the short-, medium- and long-term of material risks and opportunities arising from its impacts and dependencies on workers in the value chain.
Sustainability matters:	Working conditions (secure employment, working time, adequate wages, social dialogue, freedom of association including the existence of work councils, collective bargaining, work-life balance, health and safety), equal treatment and opportunities for all (gender equality and equal pay for work of equal value, training and skills development, the employment and inclusion of persons with disabilities, measures against violence and harassment in the workplace, diversity), other work-related rights (child labour, forced labour, adequate housing, water and sanitation, privacy)
Interaction with other ESRS:	
• **Conjunctions**	ESRS 1 (General requirements), ESRS 2 (General disclosures), ESRS S1 (Own workforce), ESRS S3 (Affected communities), ESRS S4 (Consumers and end-users)
• **Intersectional topics covered by other standards**	n/a
Interaction with other frameworks:	• ILO Declaration on Fundamental Principles and Rights at Work • International Bill of Human Rights • OECD Guidelines for Multinational Enterprises • UN Global Compact Principles • UNGPs on Business and Human Rights
Applicability:	Undertakings or groups not exceeding the average number of 750 employees may omit the information specified in the DRs of ESRS S2 for the first two years of preparation of their sustainability statement. In this case the undertaking discloses the requirements according to ESRS 2.17. (ESRS 2 X/XXII)
Overall evaluation:	

The overall evaluation contains a subjective assessment of the corresponding standard as narrative or quantitative and a count of all disclosures included in the corresponding standard. The number of disclosures was determined using professional judgement and depends on various factors. Due to the subjectivity involved, other counts may deviate.

ESRS S2 – Workers in the value chain (S2 – I/IX)

Structure and content

General disclosures

ESRS 2 SBM-2

Interests and views of stakeholders

Information on how the interests, views and rights of (actually and potentially) materially affected value chain workers, including respect for their human rights, inform the undertaking's strategy and business model.

ESRS 2 SBM-3

Material IROs and their interaction with **strategy and business model**

Whether and how IROs originate from and their relationship to strategy and business model, whether all workers in the value chain are included in the disclosure as well as disclosure of e.g. geographies at significant risk of incidents of child, forced or compulsory labour among workers in the value chain.

Impact, risk and opportunity management

S2-1 (ESRS 2 MDR-P) (VI/IX)

Policies related to value chain workers

Information on the extent to which policies address the identification, assessment, management and/or remediation of material IROs related to workers in the value chain.

S2-2 (VII/IX)

Processes for engaging with value chain workers about impacts

Information on the general processes for engaging with value chain workers and their representatives about actual and potential material impacts concerning value chain workers.

S2-3 (VIII/IX)

Processes to remediate negative impacts and **channels** for value chain workers to raise concerns

Information on the processes in place to provide for or cooperate in the remediation of negative impacts on value chain workers, as well as channels available to raise concerns and have them addressed.

S2-4 (ESRS 2 MDR-A) (IX/IX)

Actions and **resources** for managing material IROs and the effectiveness of those actions

Information on any actions and initiatives aimed at preventing, mitigating and remedying negative material impacts; achieving positive material impacts; addressing material risks and pursuing material opportunities related to workers in the value chain, including the effectiveness of those actions.

Metrics and targets

S2-5 (ESRS 2 MDR-T)

Targets related to value chain workers

Information on targets set to reduce negative impacts, advance positive impacts and/or manage material risks and opportunities related to value chain workers.

ESRS 2 MDR-M (ESRS 2 MDR-M)

Metrics related to value chain workers

Information on any metrics that the undertaking uses to evaluate performance and effectiveness in relation to material IROs.

ESRS S2 – Workers in the value chain (S2 – II/IX)

Key disclosures concerning general disclosures

Material IROs and their interaction with strategy and business model [ESRS 2 SBM-3]

- Whether all value chain workers who are likely to be materially impacted through the undertaking's own operations and the value chain are included in the scope of disclosure under ESRS 2 [11]
- Any geographies or commodities with a significant risk of child, forced or compulsory labour among value chain workers [11 (b)]
- Whether material negative impacts are either: [11 (c)]
  - Widespread or systemic in contexts where the undertaking operates or has sourcing or other business relationships or
  - Related to individual incidents or specific business relationships
- Any material risks and opportunities arising from impacts and dependencies on value chain workers [11 (e)]
- Which, if any, material risks and opportunities arising from impacts and dependencies relate to specific groups rather than all value chain workers [13]

Disclosure dependent on materiality

ESRS S2 – Workers in the value chain (S2 – III/IX)

Key disclosures concerning IRO management – 1/2

ESRS 2 MDR-P

Policies related to value chain workers [S2-1] (VI/IX)

- Whether and how policies are aligned with internationally recognised standards [19]
- Extent and nature of cases of non-respect of specific human rights policies involving value chain workers that have been reported in the value chain [19]
- Statement on the alignment with instruments such as the International Bill of Human Rights and the ILO Declaration on Fundamental Principles and Rights at Work [AR 14]

Processes to remediate negative impacts and channels for value chain workers to raise concerns [S2-3] (VIII/IX)

- Relevant and reliable data about the effectiveness of the channels to raise concerns from the perspective of value chain workers themselves [AR 26]

Disclosure dependent on materiality | Voluntary disclosures

ESRS S2 – Workers in the value chain (S2 – IV/IX)

Key disclosures concerning IRO management – 2/2

ESRS 2 MDR-A

IX/IX

Actions and resources for managing material IROs and the effectiveness of those actions [S2-4]

List of key actions taken and planned, their expected outcomes and their contribution to the achievement of policy objectives and targets [ESRS 2 MDR-A, 68 (a)]

Key actions taken (along with results) to provide remedy for those harmed by actual material impacts [ESRS 2 MDR-A, 68 (d)]

Whether the ability to implement each key action depends on specific preconditions [ESRS 2 MDR-A, 69 (a)]

Amount of current and future financial resources for the implementation of a key action or action plan [ESRS 2 MDR-A, 69 (b), (c)]

Whether and how the undertaking takes action to avoid causing or contributing to material negative impacts on value chain workers through its own practices, including, where relevant: [35]

- Procurement
- Sales
- Data use

Whether severe human rights issues and incidents connected to the value chain have been reported and disclosure of the respective incidents connected to the value chain [36]

Disclosure dependent on materiality

ESRS S2 – Workers in the value chain (S2 – V/IX)

Key disclosures concerning metrics and targets

ESRS 2 MDR-T

Targets related to value chain workers [S2-5]

- Performance against each disclosed target, including an analysis of trends or significant changes in the performance [ESRS 2 MDR-T, 80 (j)]
- Information on how each target is monitored and reviewed and whether progress is in line with what had been initially planned [ESRS 2 MDR-T, 80 (j)]

ESRS 2 MDR-M

Metrics related to value chain workers [ESRS 2 MDR-M]

- Metrics used to evaluate performance and effectiveness in relation to material IROs [ESRS 2 MDR-M, 75]
- Whether the measurement of the metric is validated by an external body other than the assurance provider and, if so, which body [ESRS 2 MDR-M, 77 (b)]

Legend: Disclosure dependent on materiality

ESRS S2 – Workers in the value chain (S2 – VI/IX)

Disclosure Requirement ESRS S2-1 | Policies related to value chain workers

Policies on material IROs related to value chain workers

- Summarised information on the key content, scope and communication of the policies as per ESRS 2 MDR-P [ESRS 2.63 ff.]
- Specification if such policies cover specific groups or all value chain workers [16]
- Whether the undertaking has a supplier code of conduct and whether it includes provisions addressing the safety of workers, precarious work, human trafficking, the use of forced or child labour and whether such provisions are fully in line with applicable ILO standards [18, AR 15]
- Whether and how policies are aligned with internationally recognised standards, including UNGPs on Business and Human Rights [19]
- Extent and nature of cases of non-respect of the UNGPs on Business and Human Rights, ILO Declaration on Fundamental Principles and Rights at Work or OECD Guidelines for Multinational Enterprises that have been reported in the value chain [19]

Human rights policy commitments relevant to value chain workers

- Description of human rights policy commitments [17]
- Description of processes and mechanisms to monitor compliance with UNGPs on Business and Human Rights, ILO Declaration on Fundamental Principles and Rights at Work or OECD Guidelines for Multinational Enterprises [17]
  - Material matters and general approach to respecting human rights, including labour rights, of value chain workers [17 (a)]
  - Material matters and general approach to engaging with value chain workers [17 (b)]
  - Material matters and general approach to measures to provide and/or enable remedy for human rights impacts [17 (c)]
- Whether policies in relation to value chain workers explicitly address trafficking in human beings, forced or compulsory labour and child labour [18]

Disclosure dependent on materiality

ESRS S2 – Workers in the value chain (S2 – VII/IX)

Disclosure Requirement ESRS S2-2 | Processes for engaging with value chain workers about impacts

Whether and how the perspectives of value chain workers are taken into account in the decision-making processes concerning impacts of the undertaking [22]

Statement on the lack of general processes for engaging with value chain workers and their representatives about actual and potential impacts [24]

Whether engagement occurs with…

- Value chain workers [22 (a)]
- Their legitimate representatives [22 (a)]
- Credible proxies with insight into the situation of the stakeholders [22 (a)]
- Specific vulnerable and/or marginalised groups (steps to gain insight into their situation) [23]
  - Women workers, migrant workers and/or workers with disabilities

Stages, type and frequency of engagement

- Stage [22 (b)]
  - Determining mitigation approaches [AR 18 (a)]
  - Evaluation of effectiveness [AR 18 (a)]
- Type [22 (b)]
  - Participation, consultation and/or information [AR 18 (b)]
- Frequency [22 (b)]
  - On a regular basis or at certain points [AR 18 (c)]
  - As a response to legal requirements and/or stakeholder requests [AR 18 (c)]
  - Integration in decision-making processes [AR 18 (c)]

Operational responsibility for engagement

- Function and the most senior role with operational responsibility for engagement, and that results inform the approach [22 (c)]
  - Consider whether the undertaking requires relevant staff to have certain skills or provides training or capacity-building [AR 18 (d)]

Agreements to respect human rights [22 (d)]

- Explanation of Global Framework Agreements
- Right to bargain collectively
- How the agreement enables the undertaking to gain insight into workers' perspectives

Assessment of effectiveness [22 (e)]

- How the undertaking assesses the effectiveness of its engagement
- Resulting agreements or outcomes

Disclosure dependent on materiality (shaded boxes) | Explanation/Examples (dashed boxes)

ESRS S2 – Workers in the value chain (S2 – VIII/IX)

Disclosure Requirement ESRS S2-3 | Processes to remediate negative impacts and channels to raise concerns

Description of processes to remediate negative impacts

- General approach to and processes for providing or contributing to remedy [27 (a)]
- Whether and how the undertaking assesses that the remedy provided is effective [27 (a)]
- Statement on lack of channels for raising concerns and/or lack of support of the availability of such channels [29]

Description of channels to raise concerns or needs

- Description of any specific channels [27 (b)]
  - *Grievance mechanisms, hotlines, trade unions, dialogue processes or other means [AR 22]*
- Whether channels are established by the undertaking itself and/or through third-party mechanisms [27 (b)]
  - *Third-party mechanisms include e.g. those operated by the government, NGOs, industry associations or other collaborative initiatives [AR 24]*
- Description of processes to support or require the availability of such channels in the workplace of value chain workers [27 (c)]
- How the undertaking tracks and monitors issues raised and addressed [27 (d)]
- How the effectiveness of channels is ensured, including through involvement of stakeholders who are the intended users [27 (d)]

Perception of individuals raising concerns and their protection

- Whether and how value chain workers' awareness and trust in the established structures and processes is assessed [28]
- Whether policies related to the protection of individuals against retaliation exist [28]

Disclosure dependent on materiality | Explanation/Examples

ESRS S2 – Workers in the value chain (S2 – IX/IX)

Disclosure Requirement ESRS S2-4 | Actions and resources for managing material IROs and the effectiveness of those actions

Impacts | **Risks and opportunities**

Summarised information on actions and resources to manage IROs as per ESRS 2 MDR-A [ESRS 2.66 ff.]

Severe human rights issues and incidents connected to the value chain [36]

Addressing material impacts

- Description of the processes to identify what action is needed and appropriate in response to a particular actual or potential negative impact on value chain workers [33 (a)]
- Description of the approach to taking action in relation to specific material negative impacts on value chain workers, including any action related to purchasing or other internal practices, capacity-building or other forms of engagement, or collaborative action with industry peers or other relevant parties [33 (b)]
- Whether and how the undertaking has taken action to provide or enable remedy for actual material impacts [32 (b)]
- Any additional actions or initiatives the undertaking has in place with the primary purpose of delivering positive impacts for value chain workers [32 (c)]
- How the undertaking tracks and assesses effectiveness of actions and initiatives in delivering intended outcomes for value chain workers [32 (d)]

Addressing material risks

- How the undertaking tracks effectiveness of actions in practice [34 (a)]
- How the undertaking ensures that processes to provide or enable remedy in the event of material negative impacts are available and effective [33 (c)]
- Whether and how the undertaking takes action to avoid causing or contributing to material negative impacts on value chain workers through its own practices, including, where relevant, practices in relation to: [35]
  - Procurement
  - Sales
  - Data use

Disclosure dependent on materiality

ESRS S3 – Affected communities

Objective:	Enabling understanding of: • The material positive and negative actual and potential impacts of the undertaking on affected communities. • Any actions taken and the result of such actions to prevent, mitigate or remediate actual or potential negative impacts and to address risks and opportunities. • The nature, type and extent of the undertaking's material risks and opportunities arising from its impacts and dependencies on affected communities and how it manages them. • The financial effects on the undertaking over the short-, medium- and long-term of material risks and opportunities arising from its impacts and dependencies on affected communities.
Sustainability matters:	Communities' economic, social and cultural rights (adequate housing, adequate food, water and sanitation, land-related and security-related impacts), communities' civil and political rights (freedom of expression, freedom of assembly, impacts on human rights defenders), rights of indigenous peoples (free, prior and informed consent, self-determination, cultural rights)
Interaction with other ESRS:	
• **Conjunctions**	ESRS 1 (General requirements), ESRS 2 (General disclosures), ESRS S1 (Own workforce), ESRS S2 (Workers in the value chain), ESRS S4 (Consumers and end-users)
• **Intersectional topics covered by other standards**	n/a
Interaction with other frameworks:	• International Bill of Human Rights • ILO Convention concerning Indigenous Peoples • ILO Declaration on Fundamental Principles and Rights at Work • OECD Guidelines for Multinational Enterprises • UN Declaration on the Rights of Indigenous Peoples • UNGPs on Business and Human Rights
Applicability:	Undertakings or groups not exceeding the average number of 750 employees may omit the information specified in the DRs of ESRS S3 for the first two years of preparation of their sustainability statement. In this case the undertaking discloses the requirements according to ESRS 2.17. (ESRS 2 X/XXII)
Overall evaluation:	 

The overall evaluation contains a subjective assessment of the corresponding standard as narrative or quantitative and a count of all disclosures included in the corresponding standard. The number of disclosures was determined using professional judgement and depends on various factors. Due to the subjectivity involved, other counts may deviate.

ESRS S3 – Affected communities (S3 – I/IX)

Structure and content

General disclosures

ESRS 2 SBM-2

Interests and views of stakeholders

Information on how the interests, views and rights of affected communities, including respect for their human rights (and their rights as indigenous peoples), inform the undertaking's strategy and business model.

ESRS 2 SBM-3

Material IROs and their interaction with **strategy and business model**

Whether and how IROs originate from and their relationship to strategy and business model, whether all affected communities are included in the disclosure as well as disclosure of e.g. the nature of negative and positive impacts.

Impact, risk and opportunity management

S3-1 (ESRS 2 MDR-P) (VI/IX)

Policies related to affected communities

Information on the extent to which policies address the identification, assessment, management and/or remediation of material IROs related to affected communities.

S3-2 (VII/IX)

Processes for engaging with affected communities about impacts

Information on the general processes for engaging with affected communities and their representatives about actual and potential material impacts concerning affected communities.

S3-3 (VIII/IX)

Processes to remediate negative impacts and **channels** for affected communities to raise concerns

Information on the processes in place to provide for or cooperate in the remediation of negative impacts on affected communities, as well as channels available to raise concerns and have them addressed.

S3-4 (ESRS 2 MDR-A) (IX/IX)

Actions and **resources** for managing material IROs and the effectiveness of those actions

Information on any actions and initiatives aimed at preventing, mitigating and remedying negative material impacts; achieving positive material impacts; addressing material risks and pursuing material opportunities related to affected communities, including the effectiveness of those actions.

Metrics and targets

S3-5 (ESRS 2 MDR-T)

Targets related to affected communities

Information on targets set to reduce negative impacts, advance positive impacts and/or manage material risks and opportunities related to affected communities.

ESRS 2 MDR-M (ESRS 2 MDR-M)

Metrics related to affected communities

Information on any metrics that the undertaking uses to evaluate performance and effectiveness in relation to material IROs.

ESRS S3 – Affected communities (S3 – II/IX)

Key disclosures concerning general disclosures

Material IROs and their interaction with strategy and business model [ESRS 2 SBM-3]

Whether all affected communities who are likely to be materially impacted through the undertaking's own operations and the value chain are included in the scope of disclosure under ESRS 2 [9]

Whether material negative impacts are either: [9 (b)]
- Widespread or systemic in contexts where the undertaking operates or has sourcing or other business relationships or
- Related to individual incidents or specific business relationships

Any material risks and opportunities arising from impacts and dependencies on affected communities [9 (d)]

Which, if any, material risks and opportunities arising from impacts and dependencies relate to specific groups rather than all affected communities [11]

Disclosure dependent on materiality

ESRS S3 – Affected communities (S3 – III/IX)

Key disclosures concerning IRO management – 1/2

Policies related to affected communities [S3-1] (ESRS 2 MDR-P) (VI/IX)

- Whether and how policies are aligned with internationally recognised standards [17]
- Extent and nature of cases of non-respect of specific human rights policies involving affected communities that have been reported in the undertaking's own operations and the value chain [17]
- Statement on the alignment with instruments such as the International Bill of Human Rights, the UN Declaration on the Rights of Indigenous Peoples and the ILO Convention concerning Indigenous Peoples [AR 10]

Processes to remediate negative impacts and channels for affected communities to raise concerns [S3-3] (VIII/IX)

- Relevant and reliable data about the effectiveness of the channels to raise concerns from the perspective of affected communities themselves [AR 23]

Disclosure dependent on materiality | Voluntary disclosure

ESRS S3 – Affected communities (S3 – IV/IX)

Key disclosures concerning IRO management – 2/2

ESRS 2 MDR-A

IX/IX

Actions and resources for managing material IROs and the effectiveness of those actions [S3-4]

- List of key actions taken and planned, their expected outcomes and their contribution to the achievement of policy objectives and targets [ESRS 2 MDR-A, 68 (a)]
- Key actions taken (along with results) to provide remedy for those harmed by actual material impacts [ESRS 2 MDR-A, 68 (d)]
- Whether the ability to implement each key action depends on specific preconditions [ESRS 2 MDR-A, 69 (a)]
- Amount of current and future financial resources for the implementation of a key action or action plan [ESRS 2 MDR-A, 69 (b), (c)]
- Whether and how the undertaking takes action to avoid causing or contributing to material negative impacts on affected communities through its own practices, including, where relevant: [35]
  - Planning
  - Land acquisition and exploitation
  - Finance
  - Extraction or production of raw materials
  - Use of natural resources
  - Management of environmental impacts
- Whether severe human rights issues and incidents connected to affected communities have been reported and disclosure of the respective incidents connected to affected communities [36]

Disclosure dependent on materiality

ESRS S3 – Affected communities (S3 – V/IX)

Key disclosures concerning metrics and targets

ESRS 2 MDR-T

Targets related to affected communities [S3-5]

- Performance against each disclosed target, including an analysis of trends or significant changes in the performance [ESRS 2 MDR-T, 80 (j)]
- Information on how each target is monitored and reviewed and whether progress is in line with what had been initially planned [ESRS 2 MDR-T, 80 (j)]

ESRS 2 MDR-M

Metrics related to affected communities [ESRS 2 MDR-M]

- Metrics used to evaluate performance and effectiveness in relation to material IROs [ESRS 2 MDR-M, 75]
- Whether the measurement of the metric is validated by an external body other than the assurance provider and, if so, which body [ESRS 2 MDR-M, 77 (b)]

Disclosure dependent on materiality

ESRS S3 – Affected communities (S3 – VI/IX)

Disclosure Requirement ESRS S3-1 | Policies related to affected communities

Policies on material IROs related to affected communities

- Summarised information on the key content, scope and communication of the policies as per ESRS 2 MDR-P [ESRS 2.63 ff.]
- Specification if such policies cover specific groups or all affected communities [14]
- Any particular policy provisions for preventing and addressing impacts on indigenous peoples [15]
- Whether and how policies are aligned with internationally recognised standards, including UNGPs on Business and Human Rights [17]
- Extent and nature of cases of non-respect of the UNGPs on Business and Human Rights, ILO Declaration on Fundamental Principles and Rights at Work or OECD Guidelines for Multinational Enterprises that have been reported in the value chain [17]

Human rights policy commitments relevant to affected communities

- Description of human rights policy commitments [16]
- Description of processes and mechanisms to monitor compliance with UNGPs on Business and Human Rights, ILO Declaration on Fundamental Principles and Rights at Work or OECD Guidelines for Multinational Enterprises [16]
    - Material matters and general approach to respecting human rights of communities and indigenous peoples specifically [16 (a)]
    - Material matters and general approach to engaging with affected communities [16 (b)]
    - Material matters and general approach to measures to provide and/or enable remedy for human rights impacts [16 (c)]

Disclosure dependent on materiality

ESRS S3 – Affected communities (S3 – VII/IX)

Disclosure Requirement ESRS S3-2 | Processes for engaging with affected communities about impacts

Whether and how the perspectives of affected communities are taken into account in the decision-making processes concerning impacts of the undertaking [21]

Statement on the lack of general processes for engaging with affected communities and their representatives about actual and potential impacts [24]

Whether engagement occurs with…
- Affected communities [21 (a)]
- Their legitimate representatives [21 (a)]
- Credible proxies with insight into the situation of the stakeholders [21 (a)]
- Specific vulnerable and/or marginalised groups (steps to gain insight into their situation) [22]
  - Women and/or girls

Stages, type and frequency of engagement
- Stage [21 (b)]
  - Determining mitigation approaches [AR 15 (a)]
  - Evaluation of effectiveness [AR 15 (a)]
- Type [21 (b)]
  - Participation, consultation and/or information [AR 15 (b)]
- Frequency [21 (b)]
  - On a regular basis or at certain points [AR 15 (c)]

Operational responsibility for engagement
- Function and the most senior role with operational responsibility for engagement, and that results inform the approach [21 (c)]
  - Consider whether the undertaking requires relevant staff to have certain skills or provides training or capacity-building [AR 15 (d)]

Agreements to respect human rights [22 (d)]
- How the undertaking assesses the effectiveness of its engagement
- Resulting agreements or outcomes

Where affected communities are indigenous peoples [23]
- How the undertaking takes into account and ensures respect of particular rights in its stakeholder engagement approach, including the right to free, prior and informed consent
- Whether and how indigenous peoples have been consulted on the mode and parameters of the engagement

Disclosure dependent on materiality (shaded boxes) | Explanation/Examples (dashed boxes)

ESRS S3 – Affected communities (S3 – VIII/IX)

Disclosure Requirement ESRS S3-3 | Processes to remediate negative impacts and channels to raise concerns

Description of processes to remediate negative impacts

- General approach to and processes for providing or contributing to remedy [27 (a)]
- Whether and how the undertaking assesses that the remedy provided is effective [27 (a)]
- Statement on lack of channels for raising concerns and/or lack of support of the availability of such channels [29]

Description of channels to raise concerns or needs

- Description of any specific channels [27 (b)]
  - Grievance mechanisms, hotlines, dialogue processes or other means [AR 18]
- Whether channels are established by the undertaking itself and/or through third-party mechanisms [27 (b)]
  - Third-party mechanisms include e.g. those operated by the government, NGOs, industry associations or other collaborative initiatives [AR 20]
- Description of processes to support or require the availability of such channels by business relationships [27 (c)]
- How the undertaking tracks and monitors issues raised and addressed [27 (d)]
- How the effectiveness of channels is ensured, including through involvement of stakeholders who are the intended users [27 (d)]

Perception of individuals raising concerns and their protection

- Whether and how affected communities' awareness and trust in the established structures and processes is assessed [28]
- Whether policies related to the protection of individuals against retaliation exist [28]

Disclosure dependent on materiality | Explanation/Examples

ESRS S3 – Affected communities (S3 – IX/IX)

Disclosure Requirement ESRS S3-4 | Actions and resources for managing material IROs and the effectiveness of those actions

Impacts	Risks and opportunities
Summarised information on actions and resources to manage IROs as per ESRS 2 MDR-A [ESRS 2.66 ff.]	
Severe human rights issues and incidents connected to affected communities [36]	

Addressing material impacts

- Description of the processes to identify what action is needed and appropriate in response to a particular actual or potential negative impact on affected communities [33 (a)]
- Description of the approach to taking action in relation to specific material negative impacts on affected communities, including any action related to the undertaking's own practices regarding land acquisition, planning and construction, operation or closure practices [33 (b)]
- Whether wider industry or collaborative action with other relevant parties will be required [33 (b)]
- Whether and how the undertaking has taken action to provide or enable remedy for actual material impacts [32 (b)]
- Any additional actions or initiatives the undertaking has in place with the primary purpose of delivering positive impacts for affected communities [32 (c)]
- How the undertaking tracks and assesses effectiveness of actions and initiatives in delivering intended outcomes for affected communities [32 (d)]

Addressing material risks

- How the undertaking tracks effectiveness of actions in practice [34 (a)]
- How the undertaking ensures that processes to provide or enable remedy in the event of material negative impacts are available and effective [33 (c)]
- Whether and how the undertaking ensures that own practices do not cause or contribute to material negative impacts on affected communities, including, where relevant, practices in relation to: [35]
  - Planning
  - Land acquisition and exploitation
  - Finance
  - Extraction or production of raw materials
  - Use of natural resources
  - Management of environmental impacts

Disclosure dependent on materiality

ESRS S4 – Consumers and end-users

Objective:

Enabling understanding of:

- The material positive and negative actual and potential impacts of the undertaking on consumers and/or end-users of its products and/or services.
- Any actions taken and the result of such actions to prevent, mitigate or remediate actual or potential negative impacts and to address risks and opportunities.
- The nature, type and extent of the undertaking's material risks and opportunities arising from its impacts and dependencies on consumers and/or end-users and how it manages them.
- The financial effects on the undertaking over the short-, medium- and long-term of material risks and opportunities arising from its impacts and dependencies on consumers and/or end-users.

Sustainability matters:

Information-related impacts (privacy, freedom of expression, access to (quality) information), personal safety (health and safety, security of a person, protection of children), social inclusion (non-discrimination, access to products and services, responsible marketing practices) of consumers and/or end-users

Interaction with other ESRS:

- **Conjunctions**

  ESRS 1 (General requirements), ESRS 2 (General disclosures), ESRS S1 (Own workforce), ESRS S2 (Workers in the value chain), ESRS S3 (Affected communities)

- **Intersectional topics covered by other standards**

  n/a

Interaction with other frameworks:

- ILO Declaration on Fundamental Principles and Rights at Work
- International Bill of Human Rights
- OECD Guidelines for Multinational Enterprises
- UN Global Compact Principles
- UNGPs on Business and Human Rights

Applicability:

Undertakings or groups not exceeding the average number of 750 employees may omit the information specified in the DRs of ESRS S4 for the first two years of preparation of their sustainability statement. In this case the undertaking discloses the requirements according to ESRS 2.17.

ESRS 2 X/XXII

Overall evaluation:

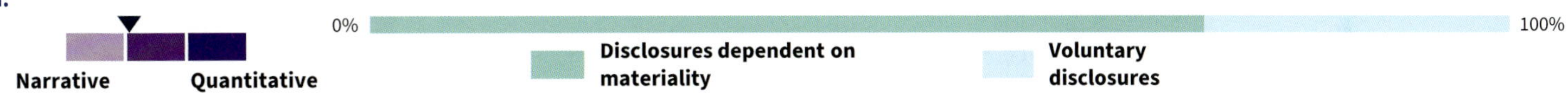

The overall evaluation contains a subjective assessment of the corresponding standard as narrative or quantitative and a count of all disclosures included in the corresponding standard. The number of disclosures was determined using professional judgement and depends on various factors. Due to the subjectivity involved, other counts may deviate.

ESRS S4 – Consumers and end-users (S4 – I/IX)

Structure and content

General disclosures

ESRS 2 SBM-2

Interests and views of stakeholders

Information on how the interests, views and rights of (actually and potentially) materially affected consumers and/or end-users, including respect for their human rights, inform the undertaking's strategy and business model.

ESRS 2 SBM-3

Material IROs and their interaction with **strategy and business model**

Whether and how IROs originate from and their relationship to strategy and business model, whether all consumers and/or end-users are included in the disclosure as well as disclosure of e.g. the nature of negative and positive impacts.

Impact, risk and opportunity management

S4-1 (ESRS 2 MDR-P) (VI/IX)

Policies related to consumers and end-users

Information on the extent to which policies address the identification, assessment, management and/or remediation of material IROs related to consumers and/or end-users.

S4-2 (VII/IX)

Processes for engaging with consumers and end-users about impacts

Information on the general processes for engaging with consumers and/or end-users and their representatives about actual and potential material impacts concerning consumers and/or end-users.

S4-3 (VIII/IX)

Processes to remediate negative impacts and **channels** for consumers and end-users to raise concerns

Information on the processes in place to provide for or cooperate in the remediation of negative impacts on consumers and/or end-users, as well as channels available to raise concerns and have them addressed.

S4-4 (ESRS 2 MDR-A) (IX/IX)

Actions and **resources** for managing material IROs and the effectiveness of those actions

Information on any actions and initiatives aimed at preventing, mitigating and remedying negative material impacts; achieving positive material impacts; addressing material risks and pursuing material opportunities related to consumers and/or end-users, including the effectiveness of those actions.

Metrics and targets

S4-5 (ESRS 2 MDR-T)

Targets related to consumers and/or end-users

Information on targets set to reduce negative impacts, advance positive impacts and/or manage material risks and opportunities related to consumers and/or end-users.

ESRS 2 MDR-M (ESRS 2 MDR-M)

Metrics related to consumers and/or end-users

Information on any metrics that the undertaking uses to evaluate performance and effectiveness in relation to material IROs.

ESRS S4 – Consumers and end-users (S4 – II/IX)

Key disclosures concerning general disclosures

Material IROs and their interaction with strategy and business model [ESRS 2 SBM-3]

- Whether all consumers and/or end-users who are likely to be materially impacted through the undertaking's own operations and the value chain are included in the scope of disclosure under ESRS 2 [10]
- Whether material negative impacts are either: [10 (b)]
  - Widespread or systemic in contexts where the undertaking sells/provides its products or services
  - Related to individual incidents or specific business relationships
- Any material risks and opportunities arising from impacts and dependencies on consumers and/or end-users [10 (d)]
- Which, if any, material risks and opportunities arising from impacts and dependencies relate to specific groups rather than all consumers and/or end-users [12]

Legend: Disclosure dependent on materiality

ESRS S4 – Consumers and end-users (S4 – III/IX)

Key disclosures concerning IRO management– 1/2

Policies related to consumers and end-users [S4-1] (ESRS 2 MDR-P) (VI/IX)

- Whether and how policies are aligned with internationally recognised standards [17]
- Extent and nature of cases of non-respect of specific human rights policies involving consumers and/or end-users that have been reported in the downstream value chain [17]
- Statement on the alignment with instruments such as the International Bill of Human Rights [AR 11]

Processes to remediate negative impacts and channels for consumers and end-users to raise concerns [S4-3] (VIII/IX)

- Relevant and reliable data about the effectiveness of the channels to raise concerns from the perspective of consumers and/or end-users themselves [AR 23]
- Number of complaints received from consumers and/or end-users during the reporting period [AR 23]

Disclosure dependent on materiality | Voluntary disclosure

ESRS S4 – Consumers and end-users (S4 – IV/IX)

Key disclosures concerning IRO management – 2/2

ESRS 2 MDR-A

IX/IX

Actions and resources for managing material IROs and the effectiveness of those actions [S4-4]

- List of key actions taken and planned, their expected outcomes and their contribution to the achievement of policy objectives and targets [ESRS 2 MDR-A, 68 (a)]
- Key actions taken (along with results) to provide remedy for those harmed by actual material impacts [ESRS 2 MDR-A, 68 (d)]
- Whether the ability to implement each key action depends on specific preconditions [ESRS 2 MDR-A, 69 (a)]
- Amount of current and future financial resources for the implementation of a key action or action plan [ESRS 2 MDR-A, 69 (b), (c)]
- Whether and how the undertaking takes action to avoid causing or contributing to material negative impacts on consumers and/or end-users through its own practices, including, where relevant: [34]
  - Marketing
  - Sales
  - Data use
- Disclosure on severe human rights issues and incidents connected to consumers and/or end-users that have been reported [35]

Disclosure dependent on materiality

ESRS S4 – Consumers and end-users (S4 – V/IX)

Key disclosures concerning metrics and targets

ESRS 2 MDR-T

Targets related to consumers and/or end-users [S4-5]

- Performance against each disclosed target, including an analysis of trends or significant changes in the performance [ESRS 2 MDR-T, 80 (j)]
- Information on how each target is monitored and reviewed and whether progress is in line with what had been initially planned [ESRS 2 MDR-T, 80 (j)]

ESRS 2 MDR-M

Metrics related to consumers and/or end-users [ESRS 2 MDR-M]

- Metrics used to evaluate performance and effectiveness in relation to material IROs [ESRS 2 MDR-M, 75]
- Whether the measurement of the metric is validated by an external body other than the assurance provider and, if so, which body [ESRS 2 MDR-M, 77 (b)]

Disclosure dependent on materiality

ESRS S4 – Consumers and end-users (S4 – VI/IX)

Disclosure Requirement ESRS S4-1 | Policies related to consumers and end-users

Policies on material IROs related to consumers and/or end-users

- Summarised information on the key content, scope and communication of the policies as per ESRS 2 MDR-P [ESRS 2.63 ff.]
- Specification if such policies cover specific groups or all consumers and/or end-users [15]
- Whether and how policies are aligned with internationally recognised standards, including UNGPs on Business and Human Rights [17]
- Extent and nature of cases of non-respect of the UNGPs on Business and Human Rights, ILO Declaration on Fundamental Principles and Rights at Work or OECD Guidelines for Multinational Enterprises that have been reported in the downstream value chain [17]

Human rights policy commitments relevant to consumers and/or end-users

- Description of human rights policy commitments [16]
- Description of processes and mechanisms to monitor compliance with UNGPs on Business and Human Rights, ILO Declaration on Fundamental Principles and Rights at Work or OECD Guidelines for Multinational Enterprises [16]
  - Material matters and general approach to respecting human rights of consumers and/or end-users [16 (a)]
  - Material matters and general approach to engaging with consumers and/or end-users [16 (b)]
  - Material matters and general approach to measures to provide and/or enable remedy for human rights impacts [16 (c)]

Disclosure dependent on materiality

ESRS S4 – Consumers and end-users (S4 – VII/IX)

Disclosure Requirement ESRS S4-2 | Processes for engaging with consumers and end-users about impacts

Whether and how the perspectives of consumers and/or end-users are taken into account in the decision-making processes concerning impacts of the undertaking [20]

Statement on the lack of general processes for engaging with consumers and/or end-users and their representatives about actual and potential material impacts [22]

Whether engagement occurs with...

- Affected consumers [20 (a)]
- Affected end-users [20 (a)]
- Their legitimate representatives [20 (a)]
- Credible proxies with insight into the situation of the stakeholders [20 (a)]
  - National consumer protection bodies [AR 14]
- Specific vulnerable and/or marginalised groups (steps to gain insight into their situation) [21]

Stages, type and frequency of engagement

- Stage [20 (b)]
  - Determining mitigation approaches [AR 16 (a)]
  - Evaluation of effectiveness [AR 16 (a)]
- Type [20 (b)]
  - Participation, consultation and/or information [AR 16 (b)]
- Frequency [20 (b)]
  - On a regular basis or at certain points [AR 16 (c)]
  - As a response to legal requirements and/or stakeholder requests [AR 16 (c)]
  - Integration in decision-making processes [AR 16 (c)]

Operational responsibility for engagement

- Function and the most senior role with operational responsibility for engagement, and that results inform the approach [20 (c)]
  - Consider whether the undertaking requires relevant staff to have certain skills or provides training or capacity-building [AR 16 (d)]

Assessment of effectiveness [20 (d)]

- How the undertaking assesses the effectiveness of its engagement
- Resulting agreements or outcomes

Disclosure dependent on materiality

Explanation/Examples

ESRS S4 – Consumers and end-users (S4 – VIII/IX)

Disclosure Requirement ESRS S4-3 | Processes to remediate negative impacts and channels to raise concerns

Description of processes to remediate negative impacts

- General approach to and processes for providing or contributing to remedy [25 (a)]
- Whether and how the undertaking assesses that the remedy provided is effective [25 (a)]
- Statement on lack of channels for raising concerns and/or lack of support of the availability of mechanisms [27]

Description of channels to raise concerns or needs

- Description of any specific channels [25 (b)]
  - *Grievance mechanisms, hotlines, dialogue processes or other means [AR 19]*
- Whether channels are established by the undertaking itself and/or through third-party mechanisms [25 (b)]
  - *Third-party mechanisms include e.g. those operated by the government, NGOs, industry associations or other collaborative initiatives [AR 21]*
- Description of processes to support or require the availability of such channels by business relationships [25 (c)]
- How the undertaking tracks and monitors issues raised and addressed [25 (d)]
- How the effectiveness of channels is ensured, including through involvement of stakeholders who are the intended users [25 (d)]

Perception of individuals raising concerns and their protection

- Whether and how consumers' and/or end-users' awareness and trust in the established structures and processes is assessed [26]
- Whether policies related to the protection of individuals against retaliation exist [26]

Disclosure dependent on materiality

Explanation/Examples

ESRS S4 – Consumers and end-users (S4 – IX/IX)

Disclosure Requirement ESRS S4-4 | Actions and resources for managing material IROs and the effectiveness of those actions

Impacts	Risks and opportunities

Summarised information on actions and resources to manage IROs as per ESRS 2 MDR-A [ESRS 2.66 ff.]

Severe human rights issues and incidents connected to consumers and/or end-users [35]

Addressing material impacts

- Description of the processes to identify what action is needed and appropriate in response to a particular actual or potential negative impact on consumers and/or end-users [32 (a)]
- Description of the approach to taking action in relation to specific material negative impacts on consumers and/or end-users, including any action related to own practices regarding: [32 (b)]
  - Product design
  - Marketing
  - Sales
- Whether wider industry or collaborative action with other relevant parties will be required [32 (b)]
- Whether and how the undertaking has taken action to provide or enable remedy for actual material impacts [31 (b)]
- Any additional actions or initiatives the undertaking has in place with the primary purpose of positively contributing to improved social outcomes for consumers and/or end-users [31 (c)]
- How the undertaking tracks and assesses effectiveness of actions and initiatives in delivering intended outcomes for consumers and/or end-users [31 (d)]

Addressing material risks

- How the undertaking tracks effectiveness of actions in practice [33 (a)]
- How the undertaking ensures that processes to provide or enable remedy in the event of material negative impacts are available and effective [32 (c)]
- Whether and how the undertaking ensures that own practices do not cause or contribute to material negative impacts on consumers and/or end-users, including, where relevant, practices in relation to: [34]
  - Marketing
  - Sales
  - Data use

Disclosure dependent on materiality

7 Governance

ESRS G1 (I/II)

Overview of sustainability matters of ESRS G-Standard

Overview of ESRS XI/XIII

G1-I/VII

Topics

Business conduct G1

Sub-topics

- Corporate culture
- Protection of whistle-blowers
- Animal welfare
- Political engagement
- Management of relationships with suppliers including payment practices
- Corruption and bribery

Sub-sub-topics

Corruption and bribery:

- Prevention and detection including training
- Incidents

ESRS G1 (II/II)

Cross-reference between ESRS 2 and ESRS G-Standard

	General disclosures						IRO management		Metrics and targets	
	Topical disclosures to be reported alongside ESRS 2 disclosures						MDRs to be applied alongside topical ESRS			
	Governance		Strategy		Processes	Transition plan	Policies	Actions and resources	Metrics	Targets
	GOV-1	GOV-3	SBM-2	SBM-3	IRO-1		MDR-P	MDR-A	MDR-M	MDR-T
ESRS Standard: **G1** Business Conduct	G1-II/VII				G1-I/VII		G1-I/VII	G1-III/VII G1-VI/VII	G1-IV/VII G1-VII/VII	G1-IV/VII

Intentionally left blank

ESRS G1 – Business conduct

Objective:	Enabling understanding of: • The strategy and approach, processes and procedures and performance of the undertaking in respect of business conduct.
Sustainability matters:	Corporate culture, protection of whistle-blowers, animal welfare, political engagement, management of relationships with suppliers including payment practices, corruption and bribery (prevention and detection including training, incidents)
Interaction with other ESRS:	
• **Conjunctions**	ESRS 1 (General requirements), ESRS 2 (General disclosures)
• **Intersectional topics covered by other standards**	n/a
Interaction with other frameworks:	• UN Convention against Corruption • Whistleblower Directive

Overall evaluation:

The overall evaluation contains a subjective assessment of the corresponding standard as narrative or quantitative and a count of all disclosures included in the corresponding standard. The number of disclosures was determined using professional judgement and depends on various factors. Due to the subjectivity involved, other counts may deviate.

ESRS G1 – Business conduct (G1 – I/VII)

Structure and content

General disclosures

ESRS 2 GOV-1

The **role of the AMSB**

Information on the role of the AMSB related to business conduct as well as the expertise of these departments regarding business conduct matters.

ESRS 2 IRO-1

Description of the **processes** to **identify** and **assess** material **IROs**

Information on all relevant criteria used in the process, including location, activity, sector and structure of the transaction.

Impact, risk and opportunity management

G1-1 (ESRS 2 MDR-P)

Business conduct **policies** and corporate culture

Information on the extent to which policies address the identification, assessment, management and/or remediation of material IROs related to business conduct as well as information on the approach to corporate culture.

G1-2

Management of relationships with suppliers

Information on the management of procurement processes including fair behaviour with suppliers.

G1-3 (ESRS 2 MDR-A) (VI/VII)

Prevention and detection of corruption and bribery

Information on the key procedures to prevent, detect and address allegations about corruption and bribery, including the trainings provided to own workers and information provided internally or to suppliers.

Metrics and targets

ESRS 2 MDR-T (ESRS 2 MDR-T)

Targets related to business conduct

Information on targets set to support business conduct-related policies and address material business conduct-related IROs.

G1-4 (ESRS 2 MDR-M) (VII/VII)

Incidents of corruption or bribery

Information on the incidents related to corruption or bribery during the reporting period and the related outcomes.

G1-5 (ESRS 2 MDR-M)

Political influence and lobbying activities

Information on the activities and commitments related to exerting political influence with political contributions, including the types and purpose of lobbying activities.

G1-6 (ESRS 2 MDR-M)

Payment practices

Information on the contractual payment terms and the performance with regard to payment, especially as to how these impact SMEs and specifically with respect to late payments to SMEs.

ESRS G1 – Business conduct (G1 – II/VII)

Key disclosures concerning general disclosures

The role of the AMSB
[ESRS 2 GOV-1]

- Expertise of the AMSB on business conduct matters
  [5 (b)]

Disclosure dependent on materiality

ESRS G1 – Business conduct (G1 – III/VII)

Key disclosures concerning IRO management

ESRS 2 MDR-A

VI/VII

Prevention and detection of corruption and bribery [G1-3]

- List of key actions taken and planned, their expected outcomes and their contribution to the achievement of policy objectives and targets [ESRS 2 MDR-A, 68 (a)]
- Key actions taken (along with results) to provide remedy for those harmed by actual material impacts [ESRS 2 MDR-A, 68 (d)]
- Whether the ability to implement each key action depends on specific preconditions [ESRS 2 MDR-A, 69 (a)]
- Amount of current and future financial resources for the implementation of a key action or action plan [ESRS 2 MDR-A, 69 (b), (c)]
- Share of functions-at-risk covered by anti-corruption and anti-bribery training [21 (b)]
- Extent to which anti-corruption and anti-bribery training is given to members of AMSB [21 (c)]

Disclosure dependent on materiality

ESRS G1 – Business conduct (G1 – IV/VII)

Key disclosures concerning metrics and targets – 1/2

ESRS 2 MDR-T

Targets related to business conduct [ESRS 2 MDR-T]

- Performance against each disclosed target, including an analysis of trends or significant changes in the performance [ESRS 2 MDR-T, 80 (j)]
- Information on how each target is monitored and reviewed and whether progress is in line with what had been initially planned [ESRS 2 MDR-T, 80 (j)]

ESRS 2 MDR-M

VII/VII

Incidents of corruption or bribery [G1-4]

- Whether the measurement of the metric is validated by an external body other than the assurance provider and, if so, which body [ESRS 2 MDR-M, 77 (b)]
- Number of convictions and amount of fines for violation of anti-corruption and anti-bribery laws [24 (a)]
- Number and nature of confirmed incidents of corruption or bribery [25 (a)]
- Number of confirmed incidents in which own workers were dismissed or disciplined for corruption or bribery-related incidents [25 (b)]
- Number of confirmed incidents relating to contracts with business partners that were terminated or not renewed due to violations related to corruption or bribery [25 (c)]

Disclosure dependent on materiality | Voluntary disclosure

ESRS G1 – Business conduct (G1 – V/VII)

Key disclosures concerning metrics and targets – 2/2

Political influence and lobbying activities [G1-5]

ESRS 2 MDR-M

- Whether the measurement of the metric is validated by an external body other than the assurance provider and, if so, which body [ESRS 2 MDR-M, 77 (b)]
- Monetary value of: [29 (b) (i)]
  - Financial political contributions made directly and indirectly
  - In-kind political contributions made directly and indirectly
    - Country or geographical area (where relevant)
    - Type of recipient/beneficiary
- Monetary amount of internal and external expenses for financial or in-kind contributions with regard to lobbying [AR 12 (a)]
- Amount paid for membership to lobbying associations [AR 12 (b)]

Payment practices [G1-6]

ESRS 2 MDR-M

- Whether the measurement of the metric is validated by an external body other than the assurance provider and, if so, which body [ESRS 2 MDR-M, 77 (b)]
- Average number of days to pay an invoice [33 (a)]
- Standard payment terms in number of days [33 (b)]
  - Main category of suppliers
- Share of payments aligned with standard payment terms [33 (b)]
- Number of legal proceedings currently outstanding for late payments [33 (c)]

Disclosure dependent on materiality | Voluntary disclosure | Breakdowns

ESRS G1 – Business conduct (G1 – VI/VII)

Disclosure Requirement ESRS G1-3 | Prevention and detection of corruption and bribery

Anti-corruption and anti-bribery system: System to prevent and detect, investigate and respond to allegations or incidents relating to corruption and bribery including the related training [16]

Summarised information on actions and resources to manage IROs as per ESRS 2 MDR-A [ESRS 2.66 ff.]

Anti-corruption and anti-bribery system

- Description of the procedures in place to prevent, detect and address allegations or incidents of corruption and bribery [18 (a)]
- Whether the investigators or investigating committee are separate from the chain of management involved in the matter [18 (b)]
- Description of the process to report outcomes to the AMSB [18 (c)]
- Statement on lack of procedures to address corruption and bribery [19]
- Plans to adopt procedures to address corruption and bribery [19]
- How the policies are communicated to ensure that they are accessible and that their implications are understood [20]
- Details about risk assessments, internal programmes, internal control procedures and communication [AR 5, AR 6]

Training

- Nature, scope and depth of anti-corruption and anti-bribery training programmes [21 (a)]
- Share of functions-at-risk covered by anti-corruption and anti-bribery training [21 (b)]
- Extent to which anti-corruption and anti-bribery training is given to members of the AMSB [21 (c)]
- Analysis of the training activities by e.g. region of training or category of own workforce where the programmes differ significantly [AR 7]

Disclosure dependent on materiality | Voluntary disclosure | Explanation/Examples

ESRS G1 – Business conduct (G1 – VII/VII)

Disclosure Requirement ESRS G1-4 | Incidents of corruption or bribery

Summarised information on metrics and the methodologies and assumptions used as per ESRS 2 MDR-M [ESRS 2.73 ff.]

- Number of convictions for violation of anti-corruption and anti-bribery laws [24 (a)]
- Amount of fines for violation of anti-corruption and anti-bribery laws [24 (a)]
- Any actions taken to address breaches in procedures and standards of anti-corruption and anti-bribery [24 (b)]
- Number of confirmed incidents of corruption or bribery [25 (a)]
- Nature of confirmed incidents of corruption or bribery [25 (a)]
- Number of confirmed incidents in which own workers were dismissed or disciplined for corruption or bribery-related incidents [25 (b)]
- Number of confirmed incidents relating to contracts with business partners that were terminated or not renewed due to violations related to corruption or bribery [25 (c)]
- Details of public legal cases regarding corruption or bribery brought against the undertaking and its own workers during the reporting period and their outcomes [25 (d)]
  - Including cases that were initiated in previous years where the outcome was only established in the current reporting period [25 (d)]

Inclusion of incidents involving actors in the value chain only where the undertaking or its employees are directly involved [26]

Disclosure dependent on materiality | Voluntary disclosure | Explanation/Examples

References to other regulations

References to other regulations

Abbreviations as per content slides	Frameworks/Regulations	ESRS Standards with reference
Accounting Directive	Directive 2013/34/EU	ESRS 1; ESRS 2
Alliance for Water Stewardship	Alliance for Water Stewardship	E3
Birds and Habitats Directives	Directive 2009/147/EC and Council Directive 92/43/EEC	E4
Circular Economy Action Plan	EU Circular Economy Action Plan	E5
CLP-Regulation	Regulation (EC) No 1272/2008 of the European Parliament and of the Council on classification, labelling and packaging of substances and mixtures (CLP)	E2
Common International Classification of Ecosystem Services	Common International Classification of Ecosystem Services	E4
Convention for Biological Diversity	Convention for Biological Diversity	E4
Council Decision on the security rules for protecting EU classified information	Council Decision 2013/488/EU	ESRS 1
Delegated Regulation on the first set of ESRS	Commission Delegated Regulation (EU) 2023/2772	Introduction
Delegated Regulation supplementing the EU Taxonomy by establishing technical screening criteria	Commission Delegated Regulation (EU) 2021/2139	E1; E4
Delegated Regulation supplementing the EU Taxonomy by specifying the content and presentation of information	Commission Delegated Regulation (EU) 2021/2178	E1
Directive on minimum wages	Directive (EU) 2022/2041	S1
EMAS	Eco-Management and Audit Scheme Regulation (EMAS)	ESRS 1

References to other regulations

Abbreviations as per content slides	Frameworks/Regulations	ESRS Standards with reference
Environmental Action Programme	Environmental Action Programme	E3
E-PRTR Regulation	Regulation (EC) No 166/2006 on the establishment of a European Pollutant Release and Transfer Register (E-PRTR)	E2
EU Biodiversity Strategy for 2030	EU Biodiversity Strategy for 2030	E4
EU Climate Law	European Climate Law	E1
EU Taxonomy	Regulation (EU) 2020/852	ESRS 1; E1
European Convention of Human Rights	European Convention of Human Rights	S1
European Pillar of Social Rights	European Pillar of Social Rights	S1
European Social Charter (revised)	The Revised European Social Charter	S1
European Waste Catalogue	Commission Decision 2000/532/EC	E5
GHG Protocol Corporate Standard	GHG Protocol Corporate Standard (Version 2004)	E1
GHG Protocol Corporate Value Chain Standard	GHG Protocol Corporate Value Chain Accounting and Reporting Standard	E1

References to other regulations

Abbreviations as per content slides	Frameworks/Regulations	ESRS Standards with reference
GRI 305	Global Reporting Initiative Standard 305: Emissions	E1
GRI Sector Standards	Global Reporting Initiative Sector Standards	ESRS 1
GRI Standards	Global Reporting Initiative Standards	ESRS 1
ILO Convention concerning Indigenous Peoples	ILO Convention concerning Indigenous and Tribal Peoples (ILO No. 169)	S3
ILO Declaration on Fundamental Principles and Rights at Work	ILO Declaration on Fundamental Principles and Rights at Work	S1; S2; S3; S4
ILO Fundamental Conventions	ILO Fundamental Conventions	S1
ILO List of Occupational Diseases	ILO List of Occupational Diseases	S1
Industrial Emissions Directive	Directive 2010/75/EU	E2
International Bill of Human Rights	International Bill of Human Rights	S1; S2; S3; S4
ISO 14064 Quantification and reporting of greenhouse gas emissions and removals	ISO 14064-1:2018 Quantification and reporting of greenhouse gas emissions and removals	E1
ISSB	International Sustainability Standards Board	ESRS 1

References to other regulations

Abbreviations as per content slides	Frameworks/Regulations	ESRS Standards with reference
IUCN Red List of Threatened Species	International Union for Conservation of Nature (IUCN) Red List of Threatened Species	E4
Kunming-Montreal Global Biodiversity Framework	Kunming-Montreal Global Biodiversity Framework	E4
Marine Strategy Framework Directive	Directive 2008/56/EC	E4; E3
Maritime Spatial Planning Directive	Directive 2014/89/EU	E3
Natura 2000 network of protected areas	Natura 2000 network of protected areas	E4
OECD Guidelines for Multinational Enterprises	OECD Guidelines for Multinational Enterprises	ESRS 1; S1; S2; S3; S4
Planetary Boundaries	Planetary Boundaries Framework	E4
Platform on Biodiversity and Ecosystem Services	Intergovernmental Science-Policy Platform on Biodiversity and Ecosystem Services	E4
REACH-Regulation	Regulation (EC) No 1907/2006 of the European Parliament and of the Council concerning the Registration, Evaluation, Authorisation and Restriction of Chemicals (REACH)	E2
Recommendation on methods for life cycle environmental performance	Commission Recommendation 2013/179/EU	E5
Regulation establishing the European Defence Fund	Regulation (EU) 2021/697	ESRS 1
Science-Based Targets Initiative for Nature	Initial Guidance for Business of the Science-Based Targets Initiative for Nature	E3; E5

References to other regulations

Abbreviations as per content slides	Frameworks/Regulations	ESRS Standards with reference
SDGs	UN Sustainable Development Goals (SDGs)	E3
Specific indicators developed by the Joint Research Center	Land-use related environmental indicators for Life Cycle Assessment by the Joint Research Center	E4
Sustainable Finance Disclosure Regulation (SFDR)	Regulation (EU) 2019/2088	ESRS 1
The Charter of Fundamental Rights of the European Union	The Charter of Fundamental Rights of the European Union	S1
UN Convention against Corruption	UN Convention against Corruption	G1
UN Convention on Rights of Persons with Disabilities	UN Convention on Rights of Persons with Disabilities	S1
UN Declaration on the Rights of Indigenous Peoples	UN Declaration on the Rights of Indigenous Peoples	S3
UN Global Compact Principles	UN Global Compact Principles	S2; S4
UNESCO World Heritage Sites	UNESCO World Heritage Sites	E4
UNGPs on Business and Human Rights	UN Guiding Principles on Business and Human Rights	ESRS 1; S1; S2; S3; S4
Waste Framework Directive	Directive 2008/98/EC	E5
Water Framework Directive	Directive 2000/60/EC	E3
Whistleblower Directive	Directive (EU) 2019/1937	G1

References to tools and databases

References to tools and databases

Tools and databases	ESRS Standards with reference
European Labour Force Survey [S1-10]	S1 – XI/XIII
Sustainable Trade Initiative [S1-10]	S1 – XI/XIII
Wage Indicator Foundation [S1-10]	S1 – XI/XIII
Fair Wage Network [S1-10]	S1 – XI/XIII
World Resources Institute's Aqueduct Water Risk Atlas [E3-4]	E3 – IX/X

Relevant definitions

Relevant definitions

Actions:	Actions refer to • Actions and action plans (including transition plans) that are undertaken to ensure that the undertaking delivers against targets set and through which the undertaking seeks to address material IROs; and • Decisions to support these with financial, human or technological resources.
Actors in the value chain:	Individuals or entities in the upstream or downstream value chain. The actor is considered downstream from the undertaking (e.g. distributors, customers) when it receives products or services from the undertaking; it is considered upstream from the undertaking (e.g. suppliers) when it provides products or services that are used in the production of the undertaking's own products or services.
Adequate wage:	A wage that provides for the satisfaction of the needs of the worker and his/her family in the light of national economic and social conditions.
Administrative, management and supervisory bodies:	The governance bodies with the highest decision-making authority in the undertaking including its committees. If in the governance structure there are no members of the administrative, management or supervisory bodies of the undertaking, the CEO and, if such function exists, the deputy CEO should be included. In some jurisdictions, governance systems consist of two tiers, where supervision and management are separated. In such cases, both tiers are included under the definition of AMSB.
Affected communities:	People or group(s) living or working in the same area that have been or may be affected by a reporting undertaking's operations or through its value chain. Affected communities can range from those living adjacent to the undertaking's operations (local communities) to those living at a distance. Affected communities include actually and potentially affected indigenous peoples.
Annual total remuneration:	Annual total remuneration to the undertaking's own workforce includes salary, bonus, stock awards, option awards, non-equity incentive plan compensation, change in pension value and nonqualified deferred compensation earnings provided over the course of a year.
Area at water risk:	A water catchment where several physical aspects related to water • Lead to one or more water bodies to be in less than good status and/or deteriorate in status (as defined in the Water Framework Directive, thus pointing to significant issues as regards water availability, quality, quantity (including high water stress); and/or • Lead to issues as regards accessibility of water, regulatory or reputational issues (including the shared use of water with communities and affordability of water) for its facilities and for the facilities of key supplier(s).
Area of high water stress:	Regions where the percentage of total water withdrawn is high (40-80%) or extremely high (greater than 80%) in the Aqueduct Water Risk Atlas. See also "water scarcity".
Biodiversity offsets:	Measures to achieve No Net Loss or a Net Gain of biodiversity for at least as long as the project's impacts are biodiversity offsets. Offsets can take the form of positive management interventions such as restoration of degraded habitat, arrested degradation or averted risk, where there is imminent or projected loss of biodiversity. (Draft ESRS E4 Biodiversity and ecosystems, EFRAG, 2022, www.efrag.org)
Biodiversity/ biological diversity:	The variability among living organisms from all sources including terrestrial, marine and other aquatic ecosystems and the ecological complexes of which they are a part. This includes variation in genetic, phenotypic, phylogenetic and functional attributes, as well as changes in abundance and distribution over time and space within and among species, biological communities and ecosystems.

Relevant definitions

Biodiversity loss:	The reduction of any aspect of biological diversity (e.g. diversity at the genetic, species and ecosystem levels) in a particular area through death (including extinction), destruction or physical/manual removal; it can refer to many scales, from global extinctions to population extinctions, resulting in decreased total diversity at the same scale.
Biodiversity sensitive area:	Natura 2000 network of protected areas, UNESCO World Heritage sites and Key Biodiversity Areas, as well as other protected areas, as referred to in Appendix D of Annex II to Delegated Regulation supplementing the EU Taxonomy by establishing technical screening criteria.
Blue economy:	The blue economy encompasses all industries and sectors related to oceans, seas and coasts, whether they are based in the marine environment (e.g. shipping, fisheries, energy generation) or on land (e.g. ports, shipyards, land-based aquaculture and algae production, coastal tourism).
Bribery:	Dishonestly persuading someone to act in your favour by giving them a gift of money or another inducement.
Business model:	The undertaking's system of transforming inputs through its business activities into outputs and outcomes that aims to fulfil the undertaking's strategic purposes and create value over the short-, medium- and long-term. ESRS use the term "business model" in the singular, although it is recognised that undertakings may have more than one business model.
Business relationships:	The relationships the undertaking has with business partners, entities in its value chain and any other non-State or State entity directly linked to its business operations, products or services. Business relationships are not limited to direct contractual relationships. They include indirect business relationships in the undertaking's value chain beyond the first tier and shareholding positions in joint ventures or investments.
Carbon credit:	A carbon credit is a convertible and transferable instrument representing GHG emissions that have been reduced, avoided or removed through projects that are verified according to recognised quality standards. Carbon credits can be issued from projects within or outside the undertaking's value chain. The carbon credits from a project should be registered in a central location where all information are publicly available. After buying the carbon credits they should be cancelled to ensure that they are not being resold or traded. (Federal Environmental Agency, 2018, www.umweltbundesamt.de)
Carbon dioxide (CO2) equivalent (eq):	The amount of carbon dioxide emission that would cause the same integrated radiative forcing or temperature change, over a given time horizon, as an emitted amount of a greenhouse gas or a mixture of GHGs (Intergovernmental Panel on Climate Change, 2023, www.ipcc.ch). CO2eq is the universal unit of measurement to indicate the global warming potential of each GHG, expressed in terms of the global warming potential of one unit of carbon dioxide. It is used to evaluate releasing (or avoiding releasing) different GHGs on a common basis.
Cascading principle:	Waste recycling and recovery in accordance with the type, quality, quantity and content of waste using multiple, consecutive material and energy recovery processes. (Federal Ministry for the Environment, Nature Conservation, Nuclear Safety and Consumer Protection, 2016, circulareconomy.europa.eu)

Relevant definitions

Circular economy:	An economic system in which the value of products, materials and other resources in the economy is maintained for as long as possible, enhancing their efficient use in production and consumption, thereby reducing the environmental impact of their use, minimising waste and the release of hazardous substances at all stages of their life cycle, including through the application of the waste hierarchy.
Circular economy principles:	The European circular economy principles are: • Usability • Reusability • Repairability • Disassembly • Remanufacturing or refurbishment • Recycling • Recirculation by the biological cycle • Other potential optimisation of product and material use.
Circular material use rate:	Recirculation of materials, components and products in practice after first use employing the following strategies (in order of preference): • Maintenance/prolonged use • Reuse/redistribution • Refurbishment/remanufacturing • Recycling, composting, or anaerobic digestion. The use rate is defined as the ratio of circular use of materials to overall use of materials.
Classified information:	EU classified information as defined in Council Decision on the security rules for protecting EU classified information or classified by one of the Member States and marked as per Appendix B of that Council Decision.
Climate change adaptation:	The process of adjustment to actual and expected climate change and its impacts.
Climate change mitigation:	The process of reducing GHG emissions and holding the increase in the global average temperature to 1.5°C above pre-industrial levels, in line with the Paris Agreement.
Climate-related physical risk:	Risks resulting from climate change that can be event-driven (acute) or from longer-term shifts (chronic) in climate patterns. Acute physical risks arise from particular hazards, especially weather-related events such as storms, floods, fires or heatwaves. Chronic physical risks arise from longer-term changes in the climate, such as temperature changes and their effects on rising sea levels, reduced water availability, biodiversity loss and changes in land and soil productivity.

Relevant definitions

Climate-related transition risk:	Risks that arise from the transition to a low-carbon and climate-resilient economy. They typically include policy risks, legal risks, technology risks, market risks and reputational risks.
Confirmed incident of corruption or bribery:	An incident of corruption or bribery that has been found to be substantiated. Confirmed incidents of corruption or bribery do not include incidents of corruption or bribery that are still under investigation at the end of the reporting period. The determination of potential non-compliance cases as substantiated may be made either by the undertaking's compliance officer or similar function or an authority. A determination as substantiated by a court of law is not required.
Consumers:	Individuals who acquire, consume or use goods and services for personal use, either for themselves or for others, and not for resale, commercial or trade, business, craft or profession purposes. Consumers include actually and potentially affected end-users.
Corporate culture:	Corporate culture expresses goals through values and beliefs. It guides the undertaking's activities through shared assumptions and group norms such as values or mission statements or a code of conduct.
Corruption:	Abuse of entrusted power for private gain, which can be instigated by individuals or organisations. It includes practices such as facilitation payments, fraud, extortion, collusion and money laundering. It also includes an offer or receipt of any gift, loan, fee, reward, or other advantage to or from any person as an inducement to do something that is dishonest, illegal, or a breach of trust in the conduct of the undertaking's business. This can include cash or in-kind benefits, such as free goods, gifts and holidays, or special personal services provided for the purpose of an improper advantage, or that can result in moral pressure to receive such an advantage.
Credible proxies:	Individuals with sufficiently deep experience in engaging with affected stakeholders from a particular region or context (for example, women workers on farms, indigenous peoples or migrant workers) who can help to effectively convey their likely concerns. In practice, this can include development and human rights NGOs, international trade unions and local civil society, including faith-based organisations.
Datapoint:	Each DR consists of one or more distinct datapoints. The term "datapoint" can refer to a narrative or a quantitative sub-element of a DR.
Discharge:	Wastewater discharge means the amount of water (in m^3) or substance (in kg biochemical oxygen demand per day or comparable) added/leached to a water body from a point or a non-point source. Sewage effluent (or discharge) means treated sewage discharged from a sewage treatment plant.
Double materiality:	Double materiality has two dimensions: impact materiality and financial materiality. A sustainability matter meets the criterion of double materiality if it is material from the impact perspective or the financial perspective or both.
Ecosystem(s):	A dynamic complex of plant, animal and micro-organism communities and their non-living environment interacting as a functional unit.
Employee:	An individual who is in an employment relationship with the undertaking according to national law or practice.

Relevant definitions

End-users:	Individuals who ultimately use or are intended to ultimately use a particular product or service for themselves.
Financial effects:	Effects from risks and opportunities that affect the undertaking's cash flow, financial position and financial performance over the short-, medium- or long-term.
Financial materiality:	A sustainability matter is material from a financial perspective if it generates risks or opportunities that affect (or could reasonably be expected to affect) the undertaking's financial position, financial performance, cash flows, access to finance or cost of capital over the short-, medium- or long-term.
Fossil fuel:	Non-renewable carbon-based energy sources such as solid fuels, natural gas and oil.
Free, prior and informed consent:	Free, prior and informed consent is a manifestation of indigenous peoples' right to self-determine their political, social, economic and cultural priorities. It constitutes three interrelated and cumulative rights of indigenous peoples: the right to be consulted; the right to participate; and the right to their lands, territories and resources. Free, prior and informed consent pertains to indigenous peoples and is recognised under international human rights law: notably the UN Declaration on the Rights of Indigenous Peoples.
Freshwater:	Groundwater and surface water with a mean annual salinity of < 0,5‰ (the limit mentioned in Annex II of the Water Framework Directive).
Functions-at-risk:	Functions deemed to be at risk of corruption and bribery as a result of their tasks and responsibilities.
GHG neutrality:	Carbon neutrality means having a balance between emitting carbon and absorbing carbon from the atmosphere in carbon sinks. (European Parliament, 2019)
GHG removal and storage:	(Anthropogenic) Removals refer to the withdrawal of GHGs from the atmosphere as a result of deliberate human activities. These include enhancing biological sinks of CO2 and using chemical engineering to achieve long-term removal and storage. Carbon capture and storage (CCS) from industrial and energy-related sources, which alone does not remove CO2 in the atmosphere, can reduce atmospheric CO2 if it is combined with bioenergy production (BECCS). Removals can be subject to reversals, which are any movement of stored GHG out of the intended storage that re-enters the surface and atmosphere.
Greenhouse Gases (GHG):	These include carbon dioxide (CO2), methane (CH4), nitrous oxide (N2O), sulphur hexafluoride (SF6), nitrogen trifluoride (NF3), hydrofluorocarbons (HFCs) and perfluorocarbons (PFCs).
Groundwater (renewable and non-renewable):	All water which is below the surface of the ground in the saturation zone and in direct contact with the ground or subsoil.

Relevant definitions

Hazardous waste: Waste which displays one or more of the hazardous properties listed in Annex III of the Waste Framework Directive.

Impact drivers: All the factors (direct or indirect drivers) that cause changes in nature, anthropogenic assets, nature's contributions to people and a good quality of life. Direct drivers have direct physical (mechanical, chemical, noise, light etc.) and behaviour-affecting impacts on nature. Indirect impact drivers operate diffusely by altering and influencing direct drivers (by affecting their level, direction or rate) as well as other indirect drivers.

Impact materiality: A sustainability matter is material from an impact perspective when it pertains to the undertaking's material actual or potential, positive or negative impacts on people or the environment over the short-, medium- or long-term. A material sustainability matter from an impact perspective includes impacts connected with the undertaking's own operations and upstream and downstream value chain, including through its products and services, as well as through its business relationships.

Impacts: The effect the undertaking has or could have on the environment and people, including effects on their human rights, connected with its own operations and upstream and downstream value chain, including through its products and services, as well as through its business relationships. The impacts can be actual or potential, negative or positive, intended or unintended and reversible or irreversible. They can arise over the short-, medium- or long-term. Impacts indicate the undertaking's contribution, negative or positive, to sustainable development.

Indigenous peoples: There is no single definition for indigenous peoples agreed on at the international level. In practice, there is convergence among international agencies on what groups can be considered indigenous peoples and should enjoy special protection as such. An important criterion for defining indigenous peoples is related to their connection to a traditional area, as defined in ILO Convention concerning Indigenous Peoples, Article 1, which states that the convention applies to:
"(a) tribal peoples in independent countries whose social, cultural and economic conditions distinguish them from other sections of the national community and whose status is regulated wholly or partially by their own customs or traditions or by special laws or regulations;
(b) peoples in independent countries who are regarded as indigenous on account of their descent from the populations which inhabited the country, or a geographical region to which the country belongs, at the time of conquest or colonisation or the establishment of present state boundaries and who, irrespective of their legal status, retain some or all of their own social, economic, cultural and political institutions".
ILO Declaration concerning Indigenous Peoples also states in Article 1 paragraph 2 that "self-identification as indigenous or tribal shall be regarded as a fundamental criterion for determining the groups to which the provisions of this Convention apply".

Relevant definitions

Internal carbon pricing scheme:	An organisational arrangement that allows an undertaking to apply carbon prices in strategic and operational decision-making. There are two types of internal carbon prices commonly used by undertakings. The first type is a shadow price, which is a theoretical cost or notional amount that the undertaking does not charge but that can be used in assessing the economic implications or trade-offs for such things as risk impacts, new investments, net present value of projects and the cost-benefit of various initiatives. The second type is an internal tax or fee, which is a carbon price charged to a business activity, product line, or other business unit based on its GHG emissions (these internal taxes or fees are similar to intracompany transfer pricing).
Key Biodiversity Area:	Sites contributing significantly to the global persistence of biodiversity, in terrestrial, freshwater and marine ecosystems. Sites qualify as global Key Biodiversity Areas if they meet one or more of 11 criteria, clustered into five categories: threatened biodiversity; geographically restricted biodiversity; ecological integrity; biological processes; and, irreplaceability.
Land-use (change):	The human use of a specific area for a certain purpose (such as residential, agriculture, recreation, industrial, etc.). Influenced by, but not synonymous with, land cover. Land-use change refers to a change in the use or management of land by humans, which may lead to a change in land cover.
Lobbying activities:	Activities carried out with the objective of influencing the formulation or implementation of policy or legislation, or the decision-making processes of governments, governmental institutions, regulators, European Union institutions, bodies, offices and agencies or standard setters. Such activities include (non-exhaustive list): • Organising or participating in meetings, conferences, events. • Contributing to/participating in public consultations, hearings or other similar initiatives. • Organising communication campaigns, platforms, networks, grassroots initiatives. • Preparing/commissioning policy and position papers, opinion polls, surveys, open letters, research work as per the activities covered by transparency register rules.
Location-based:	Location-based method quantifies Scope 2 GHG emissions based on average energy generation emission factors for defined locations, including local, subnational, or national boundaries. (GHG Protocol, “Scope 2 Guidance”, Glossary, 2015)
Marine resources:	Biological and non-biological resources found in the seas and oceans. Examples include but are not limited to deep sea minerals, gravels and seafood products.
Market-based:	Market-based method quantifies Scope 2 GHG emissions based on GHG emissions emitted by the generators from which the reporting entity contractually purchases electricity bundled with instruments, or unbundled instruments on their own. (GHG Protocol, “Scope 2 Guidance”, Glossary, 2015)
Materiality:	A sustainability matter is material if it meets the definition of impact materiality, financial materiality, or both.

Relevant definitions

Metrics:	Qualitative and quantitative indicators that the undertaking uses to measure and report on the effectiveness of the delivery of its sustainability-related policies and against its targets over time. Metrics also support the measurement of the undertaking's results in respect of affected people, the environment and the undertaking.
Microplastics:	Small pieces of plastics, usually smaller than 5mm. A growing volume of microplastics is found in the environment, including the sea and in food and drinking water. Once in the environment, microplastics do not biodegrade and tend to accumulate – unless they are specifically designed to biodegrade in the open environment. Biodegradability is a complex phenomenon, especially in the marine environment. There are increasing concerns about the presence of microplastics in different environment compartments (such as water), their impact on the environment and potentially human health.
Minimum Disclosure Requirement:	A Minimum Disclosure Requirement (MDR) sets the required content of the information that the undertaking includes when it reports on policies, actions, metrics or targets, either pursuant to a DR in an ESRS or on an entity-specific basis.
Mitigation hierarchy (biodiversity):	The mitigation hierarchy comprises: • Avoidance: measures taken to avoid creating impacts from the outset in order to completely avoid impacts on certain components of biodiversity. • Minimisation: measures taken to reduce the duration, intensity and/or extent of impacts that cannot be completely avoided, as far as is practically feasible. • Rehabilitation/restoration: measures taken to rehabilitate degraded ecosystems or restore cleared ecosystems following exposure to impacts that cannot be completely avoided and/or minimised. • Compensation or offset: measures taken to compensate for any residual significant, adverse impacts that cannot be avoided, minimised and/or rehabilitated or restored.
Net-zero target:	Setting corporate net-zero targets aligned with meeting societal climate goals means: • Reducing scope 1, 2 and 3 emissions to zero or a residual level consistent with reaching net-zero emissions at the global or sector level in eligible 1.5°C scenarios or sector pathways and • Neutralising any residual emissions at the net-zero target date – and any GHG emissions released into the atmosphere thereafter. (Science-Based Targets Corporate Net-Zero Standard, 2023, www.sciencebasedtargets.org)

Relevant definitions

Non-employees:	Non-employees in an undertaking's own workforce include both individual contractors supplying labour to the undertaking ("self-employed people") and people provided by undertakings primarily engaged in "employment activities" (NACE Code N78 (Nomenclature statistique des activités économiques dans la Communauté européenne)). Examples of contractors (self-employed persons) in the undertaking's own workforce include: • Contractors hired by the undertaking to perform work that would otherwise be carried out by an employee. • Contractors hired by the undertaking to perform work in a public area (e.g. on a road, on the street). • Contractors hired by the undertaking to deliver the work/service directly at the workplace of a client of the undertaking. Examples of people employed by a third party engaged in "employment activities" include people who perform the same work that employees carry out, such as: • People who fill in for employees who are temporarily absent (due to illness, holiday, parental leave, etc.). • People performing work additional to regular employees. • People who are dispatched temporarily from another EU member state to work for the undertaking (posted workers).
Opportunities:	Sustainability-related opportunities with positive financial effects that affect (or could reasonably be expected to affect) the undertaking's cash flows, access to finance, or cost of capital over the short-, medium- or long-term.
Own workforce:	The company's own workforce includes workers who • Are in an employment relationship with the Company ("employees") and • Non-employees who are either people with contracts with the undertaking to supply labour ("self-employed people") or people provided by undertakings primarily engaged in "employment activities" (NACE Code N78).
Packaging:	Products made of any materials of any nature to be used for the containment, protection, handling, delivery, storage, transport and presentation of goods, from raw materials to processed goods, from the producer to the user or consumer.
Pay gap:	The gap in pay between female and male employees (also known as the "gender pay gap").
Planetary Boundaries:	This concept allows to estimate a safe operating space for humanity with respect to the functioning of the Earth. The boundary level for each key Earth System process that should not be transgressed if people are to avoid unacceptable global environmental change, is quantified. The Planetary Boundaries include ocean acidification, biochemical flows, freshwater use, land-system change, biosphere integrity, climate change, novel entities, stratospheric ozone depletion and atmospheric aerosol loading.
Policy:	A set or framework of general objectives and management principles that the undertaking uses for decision-making. A policy implements the undertaking's strategy or management decisions related to a material sustainability matter. Each policy is under the responsibility of defined person(s), specifies its perimeter of application and includes one or more objectives (linked when applicable to measurable targets). A policy is validated and reviewed following the undertakings' applicable governance rules. A policy is implemented through actions or action plans.

Relevant definitions

Political contribution:	Financial or in-kind support provided directly to political parties, their elected representatives or persons seeking political office. Financial contributions can include donations, loans, sponsorships, advance payments for services, or the purchase of tickets for fundraising events and other similar practices. In-kind contributions can include advertising, use of facilities, design and printing, donation of equipment, provision of board membership, employment or consultancy work for elected politicians or candidates for office.
Pollutant:	A substance, vibration, heat, noise, light or other contaminant present in air, water or soil which may be harmful to human health and/or the environment, which may result in damage to material property, or which may impair or interfere with amenities and other legitimate uses of the environment.
Raw material:	Primary or secondary material that is used to produce a product.
Recovery:	Any operation whose principal result is waste serving a useful purpose by replacing other materials which would otherwise have been used to fulfil a particular function, or waste being prepared to fulfil that function, in the plant or in the wider economy.
Recycling:	Any recovery operation by which waste materials are reprocessed into products, materials or substances whether for the original or other purposes. It includes the reprocessing of organic material but does not include energy recovery and the reprocessing into materials that are to be used as fuels or for backfilling operations.
Renewable materials:	Material that is derived from resources that are quickly replenished by ecological cycles or agricultural processes, so that the services provided by these and other linked resources are not endangered and remain available for the next generation.
Resource inflows:	Resource that enters the organisation's infrastructure, e.g. products, (raw) materials, water and property, plant and equipment.
Resource outflows:	Resource that leaves the organisation's infrastructure, e.g. designed products and materials, packaging and waste.
Reuse:	Any operation by which products and components that are not waste are used again for the same purpose for which they were conceived. This may involve cleaning or small adjustments so the products and components are ready for the next use without significant definition.
Risks:	Sustainability related risks with negative financial effects that affect (or could reasonably be expected to affect) the undertaking's cash flows, access to finance, or cost of capital over the short-, medium- or long-term.
Scope 1 GHG emissions:	Direct GHG emissions from sources that are owned or controlled by the undertaking.
Scope 2 GHG emissions:	Indirect GHG emissions are a consequence of the operations of the undertaking but occur at sources owned or controlled by another company (GHG Protocol Corporate Value Chain Standard). Scope 2 GHG emissions are indirect emissions from the generation of purchased or acquired electricity, steam, heat, or cooling consumed by the undertaking.
Scope 3 GHG emissions:	Indirect GHG emissions are a consequence of the operations of the undertaking but occur at sources owned or controlled by another company (GHG Protocol Corporate Value Chain Standard). Scope 3 GHG emissions are all indirect emissions (not included in scope 2) that occur in the value chain of the reporting company, including both upstream and downstream emissions.

Relevant definitions

Sensitive information:	Sensitive information as defined in the Regulation establishing the European Defence Fund.
Stakeholder(s):	Those who can affect or be affected by the undertaking. There are two main groups of stakeholders: • Affected stakeholders: individuals or groups whose interests are affected or could be affected – positively or negatively – by the undertaking's activities and its direct and indirect business relationships across its value chain; and • Users of sustainability statements: primary users of general purpose financial reporting (existing and potential investors, lenders and other creditors including asset managers, credit institutions, insurance undertakings), as well as other users, including the undertaking's business partners, trade unions and social partners, civil society and non-governmental organisations, governments, analysts and academics. Some, but not all, stakeholders may belong to both groups.
Stakeholder engagement:	An ongoing process of interaction and dialogue between the undertaking and its stakeholders that enables the undertaking to hear, understand and respond to their interests and concerns.
Substances of Concern:	A substance that: • Meets the criteria laid down in Article 57 and is identified in accordance with Article 59 (1) of the REACH-Regulation; • Is classified in Part 3 of Annex VI to the CLP-Regulation in one of the following hazard classes or hazard categories: – carcinogenicity categories 1 and 2; – germ cell mutagenicity categories 1 and 2; – reproductive toxicity categories 1 and 2; – endocrine disruption for human health; – endocrine disruption for the environment; – Persistent, Mobile and Toxic or Very Persistent, Very Mobile properties; – Persistent, Bioaccumulative and Toxic or Very Persistent, Very Bioaccumulative properties; – respiratory sensitisation category 1; – skin sensitisation category 1; – chronic hazard to the aquatic environment categories 1 to 4; – hazardous to the ozone layer; – specific target organ toxicity, repeated exposure categories 1 and 2; – specific target organ toxicity, single exposure categories 1 and 2; or • Negatively affects the re-use and recycling of materials in the product in which it is present, as defined in relevant Union product-specific ecodesign requirements.

Relevant definitions

Substances of Very High Concern:	Substances that meet the criteria laid down in Article 57 of the REACH-Regulation and were identified in accordance with Article 59 (1) of that Regulation.
Supplier:	Entity upstream from the organisation (e.g. in the organisation's supply chain), which provides a product or service that is used in the development of the organisation's own products or services. A supplier can have a direct business relationship with the organisation (often referred to as a first-tier supplier) or an indirect business relationship.
Sustainability matters:	Environmental, social and human rights and governance factors, including sustainability factors defined in Article 2, point (24), of the SFDR.
Sustainability statement:	The dedicated section of the undertaking's management report where the information about sustainability matters prepared in compliance with the CSRD and the ESRS is presented.
Targets:	Measurable, outcome-oriented and time-bound goals that the undertaking aims to achieve in relation to material IROs. They may be set voluntarily by the undertaking or derive from legal requirements on the undertaking.
Transition plan:	A specific type of action plan that is adopted by the undertaking in relation to a strategic decision and that addresses: • A public policy objective; and/or • An entity-specific action plan organised as a structured set of targets and actions, associated with a key strategic decision, a major change in business model and/or particularly important actions and allocated resources.
Value chain:	The full range of activities, resources and relationships related to the undertaking's business model and the external environment in which it operates. A value chain encompasses the activities, resources and relationships the undertaking uses and relies on to create its products or services from conception to delivery, consumption and end-of-life. Relevant activities, resources and relationships include: • Those in the undertaking's own operations, such as human resources; • Those along its supply, marketing and distribution channels, such as materials and service sourcing and product and service sale and delivery; and • The financing, geographical, geopolitical and regulatory environments in which the undertaking operates. The value chain includes actors upstream and downstream from the undertaking. Actors upstream from the undertaking (e.g. suppliers) provide products or services that are used in the development of the undertaking's products or services. Entities downstream from the undertaking (e.g. distributors, customers) receive products or services from the undertaking. ESRS use the term "value chain" in the singular, although it is recognised that undertakings may have multiple value chains.
Waste:	Any substance or object which the holder discards or intends or is required to discard.
Waste hierarchy:	The waste hierarchy is the following priority order in waste prevention and management: • Prevention • Preparing for re-use • Recycling • Other recovery, e.g. energy recovery, and • Disposal

Relevant definitions

Waste management:	The collection, transport, recovery and disposal of waste, including the supervision of such operations and the after-care of disposal sites and including actions taken as a dealer or broker.
Wastewater:	Water which is of no further immediate value to the purpose for which it was used or in the pursuit of which it was produced because of its quality, quantity, or time of occurrence. Wastewater from one user can be a potential supply to a user elsewhere. Cooling water is not considered to be wastewater.
Water (recycled and reused):	Water and wastewater (treated or untreated) that has been used more than once before being discharged from the undertaking's or shared facilities' boundary, so that water demand is reduced. This may be in the same process (recycled) or in a different process within the same facility (own or shared with other undertakings) or in another of the undertaking's facilities (reused).
Water consumption:	The amount of water drawn into the boundaries of the undertaking (or facility) and not discharged back to the water environment or a third party over the course of the reporting period.
Water discharge:	The sum of effluents and other water leaving the boundaries of the organisation and released to surface water, groundwater, or third parties over the course of the reporting period.
Water scarcity:	The volumetric abundance, or lack thereof, of freshwater resources. Scarcity is human driven; it is a function of the volume of human water consumption relative to the volume of water resources in a given area. As such, an arid region with very little water, but no human water consumption would not be considered scarce, but rather arid. Water scarcity is a physical, objective reality that can be measured consistently across regions and over time. Water scarcity reflects the physical abundance of freshwater rather than whether that water is suitable for use. For instance, a region may have abundant water resources (and thus not be considered water-scarce) but have such severe pollution that those supplies are unfit for human or ecological uses.
Water withdrawal:	The sum of all water drawn into the boundaries of the undertaking from all sources for any use over the course of the reporting period.
Worker in the value chain:	An individual performing work in the value chain of the undertaking, regardless of the existence or nature of any contractual relationship with the undertaking. In the ESRS, the following is included in the scope of workers in the value chain: all workers in the undertaking's upstream and downstream value chain who are or can be materially impacted by the undertaking. This includes impacts that are connected to the undertaking's own operations and value chain, including through its products or services, as well as through its business relationships. This includes all workers who are not in the scope of the undertaking's own workforce (own workforce includes people who are in an employment relationship with the undertaking (employees) and non-employees who are either individual contractors supplying labour to the undertaking (self-employed people) or people provided by undertakings primarily engaged in employment activities (NACE Code N78)), e.g. workers of outsourced services, workers of contracted suppliers, workers for a "downstream" entity.